How to Raise Dairy Goats

Everything You Need to Know Explained Simply

Martha Maeda

HOW TO RAISE DAIRY GOATS: EVERYTHING YOU NEED TO KNOW EXPLAINED SIMPLY

Copyright © 2011 Atlantic Publishing Group, Inc.
1405 SW 6th Avenue • Ocala, Florida 34471 • Phone 800-814-1132 • Fax 352-622-1875
Website: www.atlantic-pub.com • E-mail: sales@atlantic-pub.com
SAN Number: 268-1250

Library of Congress Cataloging-in-Publication Data

Maeda, Martha, 1953-
 How to raise dairy goats : everything you need to know explained simply / by: Martha Maeda.
     p. cm.
 Includes bibliographical references and index.
 ISBN-13: 978-1-60138-378-5 (alk. paper)
 ISBN-10: 1-60138-378-9 (alk. paper)
 1. Goats. 2. Goat milk. I. Title.
 SF383.M34 2011
 636.39--dc22
                        2011011977

Printed in the United States

PROJECT MANAGER: Amy Moczynski
ASSISTANT EDITOR: Gretchen Pressley • gpressley@atlantic-pub.com
PROOFREADER: C&P Marse • bluemoon6749@bellsouth.net
INTERIOR LAYOUT: Antoinette D'Amore • addesign@videotron.ca
COVER DESIGN: Meg Buchner • meg@megbuchner.com
BACK COVER DESIGN: Jackie Miller • millerjackiej@gmail.com

Printed on Recycled Paper

A few years back we lost our beloved pet dog Bear, who was not only our best and dearest friend but also the "Vice President of Sunshine" here at Atlantic Publishing. He did not receive a salary but worked tirelessly 24 hours a day to please his parents.

Bear was a rescue dog who turned around and showered myself, my wife, Sherri, his grandparents Jean, Bob, and Nancy, and every person and animal he met (well, maybe not rabbits) with friendship and love. He made a lot of people smile every day.

We wanted you to know a portion of the profits of this book will be donated in Bear's memory to local animal shelters, parks, conservation organizations, and other individuals and nonprofit organizations in need of assistance.

– Douglas & Sherri Brown

PS: We have since adopted two more rescue dogs: first Scout, and the following year, Ginger. They were both mixed golden retrievers who needed a home.

Want to help animals and the world? Here are a dozen easy suggestions you and your family can implement today:

- *Adopt and rescue a pet from a local shelter.*
- *Support local and no-kill animal shelters.*
- *Plant a tree to honor someone you love.*
- *Be a developer — put up some birdhouses.*
- *Buy live, potted Christmas trees and replant them.*
- *Make sure you spend time with your animals each day.*
- *Save natural resources by recycling and buying recycled products.*
- *Drink tap water, or filter your own water at home.*
- *Whenever possible, limit your use of or do not use pesticides.*
- *If you eat seafood, make sustainable choices.*
- *Support your local farmers market.*
- *Get outside. Visit a park, volunteer, walk your dog, or ride your bike.*

Five years ago, Atlantic Publishing signed the Green Press Initiative. These guidelines promote environmentally friendly practices, such as using recycled stock and vegetable-based inks, avoiding waste, choosing energy-efficient resources, and promoting a no-pulping policy. We now use 100-percent recycled stock on all our books. The results: in one year, switching to post-consumer recycled stock saved 24 mature trees, 5,000 gallons of water, the equivalent of the total energy used for one home in a year, and the equivalent of the greenhouse gases from one car driven for a year.

Dedication

This book is dedicated to the people all over the world who know and
love the more than 90 million dairy goats in the world.

Table of Contents

Chapter 3: Planning for Your Dairy Goats 55

Chapter 4: Acquiring Your Herd 101

Chapter 8: Breeding Goats 175

Chapter 9: Pregnancy and Kidding...................................... 187

Chapter 10: Goat Health 213

Chapter 11: Running a Dairy Goat Business 237

he prospect of raising dairy goats conjures up images of dainty white goats grazing in fenced green pastures and buckets of rich, frothy goat's milk. As you progress through this book, you will learn that goats are indeed delightful creatures. They are also high-strung and sensitive, stubborn, curious, and quickly affected by their environment. Dairy goats, because they are being milked regularly, require more care and attention than goats kept as pets or raised for meat or fiber. Their milk must also be handled with proper care to avoid contamination and preserve its quality. Do not take on a herd of dairy goats without having a thorough understanding of their needs and requirements.

The expanding market for goat's milk, goat cheese, and candy, lotions, and soaps made with goat's milk has made commercial dairy goat farming a viable industry. If you are not prepared to undertake a commercial dairy business, you can keep your dairy goats as a source of fresh milk for your family and make occasional income by selling their kids or breeding and showing them. Dairy goats are an excellent way to diversify a small farm because they consume a variety of forage and vegetable discards and provide organic fertilizer for gardens or fields. Also, children enjoy raising and handling goats.

This book teaches all you need to know to get started with dairy goats. Learn about dairy goat breeds and how to select the best goats for your herd; how to construct fences, shelters, feeding, and milking facilities; how to feed and care for your goats; and how to recognize and deal with health problems. You will also read about goat's milk, how you can enhance the production of milk, and how the feed you give your goats affects its quality. Dairy goats must be bred often to keep up their milk production. Learn how to breed your goats and care for their kids. The last chapter covers what you must know to start a commercial dairy goat business.

I have yet to meet a single dairy goat farmer who regretted raising goats. Some have sold their goats and retired or given up dairy goat farming for one reason or another, but no one has said he or she was sorry for having embarked on the adventure. Everyone shares treasured memories of favorite goats, funny experiences, and personal accomplishments. Join them in the rewarding experience of raising, caring for, and loving dairy goats of your own.

Photo courtesy of Soggy Bottom Farms.

Chapter 1

Welcome to the World of Goats

oats have been companions to man almost since the beginning of human civilization. The goat was the second animal to be tamed, after the dog, and the first herbivore to be domesticated. Early hunter-gatherers drank goat's milk and soon learned to make cheese. Archaeologists have found evidence that goats were being kept, rather than hunted, around 8000 B.C. at Ganj Dareh, a Neolithic village in western Iran. Mounds of domestic goat bones unearthed at Jericho have been carbon-dated to 7000 – 6000 B.C.

DNA findings released in 2001 suggest goats may have been domesticated in other regions of the world as well. Gordon Luikart of the Université Joseph Fourier in Grenoble, France, and his colleagues speculate that another goat strain found in the Indian subcontinent, Mongolia, and Southeast Asia is descended from a she-goat tamed about 9,000 years ago in an area of Pakistan called Baluchistan, in the Indus Valley. Archaeological evidence suggests that Baluchistan was a major center of domestication. A third unrelated strain of yet-undetermined origin is believed to have been the ancestor of today's Swiss, Mongolian, and Slovenian breeds.

Domestic goats are thought to have shaped the development of human civilization by providing a portable supply of meat and milk that allowed hunter-gatherers to become agricultural nomads. Goats were small, easy to handle, and able to thrive in arid, semi-tropical and mountainous areas where horses, cattle, and other larger herbivores could not survive. Their skins supplied leather and pelts for robes and rugs. The woolly fur of goats was woven into the earliest cloth, and goat horns were made into drinking vessels, ornaments, and musical instruments. Goats carried their owners' belongings on their backs or pulled them on sledges as they traveled from region to region. Young kids were used as sacrificial animals for religious rites. Goats achieved mythical stature as fertility gods; the Greek god of forests and flocks, Pan, was born as a goat; and the Teutons believed that the chariot of Thor, the thunder god, was pulled through the heavens by two he-goats.

Domesticated goats spread east from the Fertile Crescent across Europe and eventually into Great Britain, sometimes escaping and establishing feral populations in remote areas where they can still be found. Today, goats are the third most plentiful animal in the world. The estimated world population of goats today is about 460 million, the majority of them in developing countries where goat products are common and widely valued.

More than half of the world's population drinks goat milk, although in the United States goat's milk is only slowly gaining in popularity. Europeans have appreciated the special attributes of goat's milk for generations, and products such as chèvre (goat's milk cheese), butter, yogurt, and ice cream made from goat's milk are popular in European nations. Until about 400 years ago, goats surpassed cattle in Europe as the preferred milking animal, and in much of the world, they still do. The reasons are simple: Goats are less expensive to purchase than larger animals, and the amount of feed a goat requires per gallon of milk output is less than that needed for cattle. Goats are hardy, smaller, and more manageable. Goats multiply rapidly. Three or four goats are a much safer investment than a single cow, because if one goat fails to produce or becomes sick, the others can replace it. These qualities make goats ideal for small dairy operations or for household use. Cow's milk only began to predominate when modern dairies became a large organized industry because larger animals could give greater quantities of milk.

Dairy Goats in the United States

Goats were brought to the Americas in the 1500s by Spanish seafarers, who carried them on their ships to provide milk and meat, and often released them on small islands where they multiplied and could be caught and slaughtered for fresh meat by future voyagers.

Angora goats

A visitor to Plymouth Colony in September 1623 noted that the colony possessed six goats. In 1849, North America's first purebred goats, seven Angora does and two bucks, were imported to South Carolina. In 1904, the first North American dairy goat exhibition was held at the St. Louis World's Fair, and in the same year, the American Milk Goat Record, now the American Dairy Goat Association (ADGA), was established. During the early 1900s, several breeds of dairy goat were imported. The first officially documented Pygmy goats were imported to the U.S. during the 1950s as zoo animals.

During the first half of the 1900s, goat's milk in the United States was marketed primarily through pharmacies as an alternative for people who were allergic to cow's milk. In the 1970s, a movement toward self-sufficiency and sustainable agriculture revived an interest in raising dairy goats. Small and adaptable, and consuming relatively few resources, one or two goats could supply all the milk a family needed. During the 1980s, Americans became increasingly interested in healthful and natural foods, and during the 1990s, the rising popularity of specialty cheeses and ethnic cuisines contributed to a growth in demand for goat milk and goat's milk products.

Before the late 20th century, travelers in rural America occasionally spotted a few goats mixed in with a farmer's livestock in the fields, a nostalgic reminder of the "old world" of European ancestors. During the last few decades, there has been an upsurge in the number of Americans raising goats for meat and milk and a steady annual increase in the number of dairy goats. According to the USDA's National Agricultural Statistics Service (NASS), there were 355,000 milk goats listed in the United States in January 2010, an increase of 6 percent over the previous year. Dairy goats are found in every state in the United States, but the largest numbers reside in Wisconsin (46,000 head), California (38,000 head), Iowa (29,500 head), and Texas (20,000 head). Because these statistics are based on dairy operations large enough to be licensed and regulated as working dairy businesses by the U.S. government, they do not include goats raised on hobby farms or kept in backyards. The population of producing dairy goats in the United States could be 400,000 or more if these smaller operations are taken into account.

The main dairy goat breeds in the United are Alpine, LaMancha, Nubian, Oberhasli, Saanen, and Toggenburg. Each of these breeds is capable of producing more than 2,000 pounds of milk per year, but the United States imports more than 50 percent of the dairy goat cheese products it consumes, mostly from France.

CASE STUDY:
MEYENBERG
GOAT MILK PRODUCTS

In 1934, Harold Jackson bought Meyenberg Goat Milk Products (**www. meyenberg.com**) because his infant son, Robert, suffered from cow milk allergies and digestive problems. Initially, Jackson Mitchell's Meyenberg Goat Milk Products produced only evaporated goat's milk, sold exclusively in pharmacies as an alternative for infants sensitive to cow milk. Robert D. Jackson assumed ownership of the company in 1954. Jackson and his wife of 43 years, Carol, turned the small evaporated milk company into the largest manufacturer of goat's milk products in the U.S. and the pre-eminent global manufacturer of goat's milk products. Recognizing competition from the proliferation of new infant formulas, Jackson began marketing goat's milk to a larger population, focusing special attention on health conscious consumers, seniors and children.

During the 1980s, a growing national awareness of health and nutrition issues renewed consumer interest in unique natural products. To meet demand, Meyenberg opened plants in Arkansas and California, doubling goat's milk production and annual sales within a year. A unique ultrapasteurization process allowed for a wider distribution of fresh goat's milk products to markets across the country, including Meyenberg ultra-pasteurized fresh whole goat milk and 1 percent low-fat pasteurized milk, evaporated, and powdered goat's milk. In 2004, the company introduced European-style goat milk butter and goat milk cheddar and, in 2008, goat cream cheese.

Recent trends have fueled yet another boom in goat's milk consumption: consumer desire to avoid foods containing additives and bovine growth hormones (BGH), a mushrooming demand for foods that are 100 percent natural, and a growing appreciation by gourmet chefs, who use goat milk to enhance recipes. Meyenberg products are currently available in 90 percent of U.S. supermarket chains, as well as through leading health-food distributors and stores. Its farms and facilities produce more than 20 million pounds of goat milk annually, and its lines of distribution stretch around the world.

Why Dairy Goats?

Goats are an ideal livestock for a small family farm operation. They are friendly and social animals compared to sheep and cattle, and their small size allows them to be managed more easily than dairy cattle, which weigh as much as ten times more. Apart from their large size, several breeds of dairy cattle are aggressive and present difficulties when the time comes to load an animal into a truck, coax it into a pen, or halter it for leading. Older children can physically manage a goat easily after some instruction and initial supervision. Goat care and ownership can open up an exciting world of showing and breeding for children.

The novice goat owner will find that dairy goats are extremely versatile. Depending upon the breed or breeds of goat you have, you can supplement your income by selling raw milk and milk products (such as soaps, lotions, and cheeses); sell wool, horns, hide, and male offspring; obtain prizes from showing and breeding; and sell goat manure for fertilizer. You can also make money by keeping a buck for breeding and charging stud fees.

Some breeds of goats are best for milk production; other breeds are prized for their meat (called chevon), or for their long hair,  which is spun into yarn. Many farmers choose to raise dairy goats because they want to develop emotional attachments to their livestock. Although dairy goats are occasionally slaughtered or sold, the nature of the dairy business is different from that of a meat goat farm, where individual goats are regularly and quickly disposed of for profit. Dairy goats are productive milkers for seven to eight years, and even longer in some cases. Although many people love the business of wool goats, which are also slaughtered less frequently, processing wool is hard work and takes practice. Marketing specialized wool is a particular challenge in the United States compared to the marketing of other goat products such as milk and meat. Raising dairy goats allows the human caretakers to bond with their livestock and, at the same time, has more potential to earn a profit.

There are several reasons why the number of dairy goat establishments is steadily increasing:

The market for goat's milk products is expanding

This is an exciting time to become involved in the dairy goat industry in the United States because the market for goat's milk products is steadily growing and is predicted to continue growing for decades to come. Goat's milk is seeing a resurgence of popularity in the United States. During the 1980s, a surplus of cow's milk led dairy farming associations to invest in a massive campaign advertising the nutritional value of milk. Doctors have observed that patients with allergies to cow's milk can often digest goat's milk, though scientific studies have not yet established the reason for this.

In recent decades, immigrants from Africa and the Middle East have established large communities in the United States. Their religious and cultural traditions value the goat highly for its milk, meat, and even for ritual slaughter as part of religious observation, opening up profitable new markets for the dairy goat entrepreneur wise enough to target them. Television exposure on food and cooking shows and the opening of new types of restaurants has increased the popularity of ethnic and nouvelle cuisines incorporating goat's milk and goat cheese. Goat meat is also gaining a reputation with health-conscious meat consumers because it has significantly less fat than similarly prepared beef, with a comparable protein content. The USDA also reports that cooked goat meat has 40 percent less saturated fat than chicken, even with the chicken skin removed.

Compared to other livestock, goats are adaptable and have only a few basic needs

Dairy goats can be raised in a relatively small area, even on an average-sized urban lot. They can also be raised indoors, though it is good for them to have space to roam outdoors. Does do not have a strong odor and are dainty and fastidious. Goats do not have to be pastured and can be fed entirely with hay and grain.

They are relatively inexpensive to purchase and do not require as much feed as cattle.

Self-sufficiency: A family can supply all its dairy needs with two or three goats

A standard dairy goat gives an average of 6 to 8 pounds (3 to 4 quarts) of milk every day. A doe must give birth to a kid before you can start milking her, but the milking can then continue throughout the year for as long as three years until the mother is bred again and is in the last two months of her pregnancy. By staggering their pregnancies, you can ensure a steady supply of milk for your family with just two or three does. A family can supply all its milk and dairy needs with a small herd of dairy goats. Goat's milk is enjoyed fresh, but it is also suitable for making sour milk, cheese, butter, yogurt, and ice cream. When you produce your own milk products, you know their origin and can be sure they do not contain hormones, antibiotics, or additives.

Goats are "green cows"

Goats require less space, cost less to feed, and produce more milk per pound of feed than cows. They also produce less methane gas; 85 million tons of methane, a primary contributor to human-induced global warming, is produced in the digestive tracts of agricultural livestock every year.

Goat farming complements organic gardening well. Goat droppings contain all the nutrients for optimal growth of plants and can be used in nearly any type of garden to fertilize flowers, herbs, vegetables, and fruit trees. A goat deposits 2 to 6 percent of its body weight every day in pellets and 1 to 4 percent of its body weight in urine. Much of this can be recovered from thick bedding on the floor of the goat shelter. The urine absorbed by this bedding contains concentrated nitrogen and potassium. Soiled bedding can amount to as much as 1.5 tons per animal every year and has an estimated N-P-K (nitrogen-phosphorus-potassium) value of 1.3-1.5-0.04. Goat manure can be composted and used as mulch. The hard dry pellets are easier to spread than cow or horse manure and do not burn plants.

Plants used for crop rotation that do not produce direct yield, such as clover, are excellent feed for goats, along with many types of garden debris. *(Not all garden debris is suitable for feeding goats. See Chapter 5.)* Goats can browse in company with sheep or cattle because they eat different plants; they prefer to browse on brush, leaves, and rough plants. Goats also browse in small areas and along the edges of roadsides.

Goats are friendly, intelligent, cheerful, and curious

Most goat keepers will testify that their goats have interesting and entertaining personalities and that they make good companions. If handled frequently, goats are easy to manage. They can be trained to walk on a leash and to obey routine commands. Goats are frequently kept as pets, but they should not be played with like dogs because they will develop bad behaviors and become unruly adults. Goats are social animals and are happier if they are kept with at least one other goat. A trained goat can be taken as a therapy animal or visitor to schools, hospitals, and senior citizens' homes to educate, entertain, and raise morale.

You can earn supplementary income by selling the extra kids and providing buck service

Because you must breed dairy goats in order for them to produce milk, you will eventually end up with surplus kids and unwanted bucklings, which you can sell for meat or to other dairy goat farmers. Some dairies breed their does to Boer and Kiko bucks to deliberately produce kids suitable for meat goats. If you keep your own buck, you can earn extra income by providing buck service (breeding services) to other herds of dairy goats.

Goats can provide additional services such as destroying noxious weeds and acting as pack animals

Goats help to maintain biodiversity by controlling the spread of invasive plants such as kudzu and blackberry. They eliminate the

need for herbicides and are useful for cleaning up hard-to-reach areas such as steep slopes. Goats also clear scrubland by eating small saplings and brush. Some goat keepers hire out their goats for these purposes, but dairy goat farmers should avoid letting their goats graze in rough areas where their udders might become scratched or injured.

Starting at the age of eight or ten months, a goat can be trained to carry a pannier or to pull a cart or plow in harness. Sure-footed and strong, a trained goat is a useful companion on a backpacking trip. A lactating doe provides the additional benefit of fresh milk on the trail. Children enjoy riding in goat carts at parties and in parades.

Why You Might Not Want to Invest in Dairy Goats

Although the prospect of raising dairy goats sounds exciting, you should consider several important things before you begin committing your time and money. First, you must be clear about your expectations. Will your dairy goats be an enjoyable hobby and a source of milk and dairy products for your immediate family, or do you hope to establish a profitable dairy farm business? How much money and time can you afford to invest in your dairy goats before they begin producing income? The dairy goat business is in its infancy in the United States. Year by year it becomes easier and more profitable as American consumers learn more about goat products and as more resources for goat owners become available. The owners of several successful large-scale goat dairies around the United States agree that it is a tough business but that it is a growing industry with a very bright future.

There is a big difference between farming dairy goats for your personal consumption and raising dairy goats as a business. In order to run a successful dairy goat business, you have to find a regular and reliable buyer for your milk or milk products. Many farmers have invested in constructing a full-fledged dairy facility, only to have their market fizzle out. If you are uncertain about your business prospects, start with two or three dairy goats as a hobby. Take time to learn about goats and get a feel for whether you want to run a full-scale dairy. As you become acquainted with other goat keepers in your area, you will grow familiar with local markets for goat's milk products, goat meat, and other services.

Dairy goats require constant care and attention

In order to produce to their full potential, dairy goats must be properly housed and fed, and lactating goats must be milked regularly at 12-hour intervals. Failure to do so will result in health problems and a decline in milk production. The milking parlor (milking room) and equipment must be kept clean and sanitary and the milk must be handled correctly. In addition, does must be bred at regular intervals and carefully observed and monitored. You will need to devote a minimum of 30 minutes each morning and evening, including weekends, vacations, and holidays, to caring for your dairy goats. If you and your family are not prepared to make this commitment, consider raising goats for meat and fiber instead. You will also need to find and train a competent assistant, such as a friend or a neighbor, who can fill in for you in case of an emergency or absence.

Your knowledge, previous experience, and available resources

If you are a complete novice, you will need assistance from other goat keepers, a veterinarian, and local agriculture officials. In some areas, there is an established community of dairy goat farmers, a local dairy goat association, and agricultural advisers who will be able to guide you. In other areas, you may be the only dairy goat farmer for miles around. You will need to establish a support network by attending shows and sales, contacting other dairy goat farmers, and finding a veterinarian who has experience with dairy goats. The farmer who sells you your dairy goats may be willing to mentor you until you become experienced. If you are inexperienced, you might start by buying a mature, lactating doe and her kid. If you know what you are doing, you might prefer to raise two or more kids and then breed them. The nature and size of your operation will also depend on the amount of land you have and the buildings existing on it. The initial cost of setting up a dairy goat operation will be much greater if you have to construct new buildings and shelters. The availability of pasture and natural forage will make feeding the goats less expensive than if you have to buy all their feed.

Local and state regulations regarding the sale of milk and dairy products

Before you sell any goat milk products for human consumption, such as milk, cheese, butter, or yogurt, research the laws in your state regarding sale of such items. Consult your local agricultural extension office, your state's dairy division, or the American Dairy Goat Association (**www.adga.org**). Depending upon the rules in the state in which you live, you may have to be licensed as a Grade A dairy in order to sell dairy products for human consumption. Many states prohibit the sale of raw milk (milk that has not been pasteurized) for human consumption from any kind of dairy. To be licensed as a Grade A dairy, you must meet strict standards that vary from state to state. These standards generally include provision of specific types of lighting; regulation of the quality of flooring in your barn; cleanliness of barn environment; types of milking and processing equipment; location of human toileting area; methods of storage and processing; and provision of a separate room for storage and cooling of milk. A large dairy operation requires a processing plant nearby that is willing to accept your milk. *Licensing will be discussed further in Chapter 11.*

If you cannot obtain a Grade A dairy license, you will only be able to make personal use of your milk. You can feed the excess to your farm animals or sell it to neighbors who are bottle-feeding lambs, kids, puppies, or kittens. You can offset your startup costs by marketing nonfood items such as goat's milk soaps and lotions, by showing your goats, or by breeding for sale.

Your neighbors' attitudes

Before you bring a goat home, find out the local regulations about keeping livestock. Some urban areas do not allow goats even if the lots are large. Goats have a tendency to escape if there is a weak spot in their enclosure, and they like to eat rose bushes, shrubs, and young trees. You do not want to turn your neighbors into enemies. Make sure your neighbors do not own vicious dogs that might escape and attack your goats, and that there are no packs of wild dogs roaming the area.

Some neighbors do not appreciate the joyful bleating noises of a herd of goats. Let your neighbors know about what you are doing and gain their approval. If they are adamantly opposed to your pasturing goats in a field adjoining their property, you will have to decide your priorities — dairy goat farm or friendly relations with your neighbors.

TIP: Good fences

The old adage that "good fences make good neighbors" is especially true for goat farmers. Because goats are quick to take advantage of any opportunity to escape and browse on choice landscaping, it is important they be kept in secure enclosures to avoid unpleasant incidents that might create bad feeling among neighbors.

Most dairy goat owners start out by purchasing just a few goats to provide milk for the family while they learn the finer points of milking and upkeep of a herd. It is important to give yourself this time to dabble in the arts of cheese making and producing soaps and lotions from goat's milk. Over time, you can learn what quantity to expect from your goats (this varies according to breed and among individuals as well), and which products you wish to concentrate upon in the future. Before you decide to run a full dairy operation, make sure you can meet all the legal requirements.

The unique nature of a goat

People who have raised goats describe them as affectionate, wily, funny, smart, strong-willed (this is usually not expressed in positive terms), sure-footed, social, easy-going, independent, and

curious. Goats are famous escape artists; if there is a way to get out of an enclosure, your goats will find it. Goats like to be up high where they can survey the world. You may have seen photos of goats standing on cars, goats in trees, and goats on roofs. These photos are not staged.

Goats appear to be curious. They observe human beings and unfamiliar circumstances with alert attention, and they are quick to assess novel situations and take advantage of them. When faced with danger, goats respond courageously, typically making a few backward jumps to observe the threat from a greater distance. Instead of fleeing blindly like a flock of sheep would, a goat will stand up to a threat. For this reason, goat bucks are often used as lead animals for herds of sheep.

Even mature goats are playful at times. Goats often seem to have a special capacity for silly or malicious behavior and can be obstinate. If a goat does not want to cooperate with the goat keeper, it will plant its feet and refuse to budge or even to eat. Efforts to force a goat to do something against its will are usually unsuccessful; it is better to try to outwit the goat.

Sheep and cattle are grazers and nibble on plants low to the ground. Goats, like deer, are browsers. They look for things growing up higher, such as brush, bushes, and the green shoots, leaves, and bark of trees. Keep this trait in mind when you consider where on your land you are going to turn your goats loose to forage. It is a myth that a goat will eat anything, including tin cans. Goats are picky eaters and select only their favorite morsels from the feed given to them; they will not eat tin cans, but they will eat the paper labels off them. A goat's tendency to explore with its lips while browsing sometimes leads observers to believe the animal is trying to eat everything in sight, when in fact it is simply curious. Once in a while, like any livestock animal, a goat may eat something unsafe such as a nail, a piece of plastic or tin, or a length of twine. Avoid injuries and expensive veterinary bills by keeping all of these dangerous items out of the reach of your livestock.

Goats are herd animals and are not happy living alone. If you have ever spent time with sheep, you might have noticed how closely together they flock. Usually a herd of a dozen will be contained within 20 square yards as they graze. Goats do not flock together quite as closely as sheep. Although they like to be within sight of one another, an individual will occasionally wander off, only to burst into a bout of panicked bleating when it discovers the herd has moved on.

Goats within a herd establish a firm hierarchy, with a lead, or alpha, buck and a herd queen who is the real leader. The duty of the alpha buck is to protect the herd from predators, breed his does, and maintain discipline. The herd typically moves in single file with the alpha buck guarding the rear when it senses danger. The less dominant males in the herd are not permitted to breed. If the alpha buck is killed, removed from the herd, or deposed by a younger and stronger buck, the herd easily accepts a new alpha male. When new goats join the herd, the hierarchy is re-established by dueling. The males fight by pressing their foreheads against each other and rarely injure themselves. (If a predator attacks, on the other hand, a goat will attempt to jab the predator's tender areas with the sharp tips of its horns.) When a herd has no males, the goat keeper typically takes on the role of alpha male.

The herd queen leads the herd and decides when and where all the goats will graze. She tastes the plants before the other goats eat, and if she eats something unpleasant, she makes a big production of spitting it out and shaking her head. The queen and her daughters get the best sleeping spot and the best spots in front of the feeding trough. The herd queen remains in her role until she becomes infirm or dies, and then the herd selects a new queen. Whenever a new doe is introduced to the herd, there will probably be some sparring to establish dominance, so it is important to separate the new doe from the others in a way that allows them to sniff and interact with each other without coming into physical contact.

TIP: A lone goat is a naughty goat

One Missouri woman who acquired a pet goat experienced some problems when he not only climbed her truck, but also ate the windshield wipers and then pretty much destroyed the vehicle. She ended up donating the truck permanently for his entertainment. She made two mistakes: first, she allowed him access to the truck; but second and more importantly, she had only one goat. Goats are social beings meant to live together in a herd. A lone goat is an insecure goat, and an insecure goat will find ways to make life more interesting and assert its dominance over its environment.

As a general rule, if you have a goat with apparent behavioral problems, ask yourself if the goat is too isolated from other goats, and look for signs of illness or disease. Check for signs that external parasites, such as flies, or internal parasites, such as worms, might be bothering the goat. Consider whether there could be some problem with feed or water quality or if the bad behavior could be related to sexual frustration. Finally, your goat might be responding to an unfamiliar environmental influence, such as being moved recently from another farm, or having problems adjusting to another new goat in the herd.

Goats can be especially prone to developing pneumonia, and most goats do not like to be wet. If a storm is approaching, you will find them already inside. Particular breeds can also be intolerant of hot weather. It is important in warmer months that your goats always have access to shade and fresh water.

Goats like to be where people are. Goat owners speak of walking miles with a half dozen goats following along behind. Some even walk goats on leashes. Goats will cuddle and will even vocalize to you. A doe will often allow strangers to pick up and nuzzle her kids without showing much concern.

Before you acquire your first goat, it is a good idea to get some experience with goat behavior by helping out at a dairy farm as a volunteer (most farmers will gladly accept your help). By doing so, you become accustomed to being around the goats, the goat operation environment, and the way goats relate to one another and to humans.

Chapter 2

Anatomy of a Dairy Goat

here are approximately 100 documented breeds and varieties of domestic goat. All goats share similar characteristics, breeding traits and flocking instincts. Goats are small ruminants, meaning their digestive systems consist of multiple stomachs. Domestic goats belong to the Bovidae family, along with cattle and other hollow-horned, cloven-hoofed ruminants, and to the Caprinae subfamily along with sheep. Domestic goats belong to Capra hircus, one of six species of goats (the other five species are wild). Domestic goats are typically categorized according to their usage as dairy, meat, or fiber-producing goats. The term "dual purpose" refers to a breed or individual goat that belongs to more than one of these categories. For example, you might have a milk-producing goat with lovely long hair that can be sheared and spun into yarn.

Goats are herbivores. Their bodies process grains, plants, and their own mothers' milk to supply all the nutrients they need. All herbivores share some common traits. For example, if you look at the head of a goat, sheep, horse, or cow, you will notice the eyes are set on the side of the head, not at the front like human eyes or those of your dog or your cat. This is because in the animal world, herbivores are most often the prey of carnivores. Carnivores have eyes that see ahead for tracking and hunting. Herbivores are stalked and set upon as they graze, and eyes in the sides of their heads allow them to see danger approaching from the side or behind. Herbivores feel safer in numbers (herds) and are almost always in a paranoid state of mind, ready to fight or flee from danger. This is the way your goats view the world.

When you speak to a veterinarian, another goat owner, or perhaps to a judge at a livestock show, they will use anatomical terms specific to a goat. The illustration below shows the main parts of a female goat's anatomy:

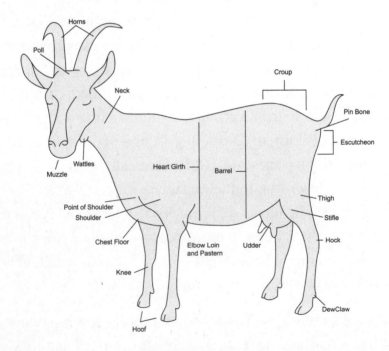

Anatomy of a goat

Digestive System

A goat is a ruminant, an animal that chews its cud. Like cattle and sheep, goats have four stomach-like compartments in their digestive systems. Each of the four — the reticulum, rumen, omasum, and abomasum — plays a unique part in digesting food matter. It is important to understand how the digestive system of a goat works so you can avoid digestive problems caused by improper feeding.

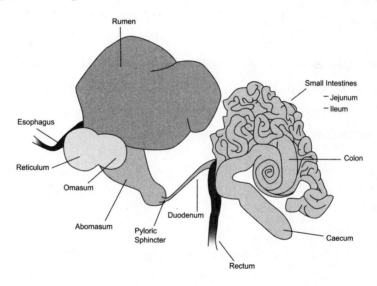

Digestive system of a goat

1. **Rumen:** This is the largest compartment, comprising about 80 percent of the whole stomach. It is sometimes called the fermentation vat because this is where most of the fermentation takes place. The rumen is most important for the digestion of fiber, and fiber makes up a large part of the goat's diet. The rumen works by muscular contractions and enzymes supplied by microorganisms living in the rumen (bacteria, fungi, and protozoa) that break down fiber and convert cellulose into volatile fatty acids that supply

75 percent of the goat's energy. The microorganisms also manufacture protein from simple nitrogen compounds in the goat's feed, as well as amino acids and all the K and B vitamins the goat needs. The walls of the rumen do not secrete any enzymes.

2. **Reticulum:** This is separated from the rumen only by a partial wall. When dissected, the inside is made up of many little cups that look like a honeycomb, so it is often called the "honeycomb" compartment. It functions as a fluid pump. Hardware and other objects a goat may consume become trapped here and can cause something called "hardware disease" in which sharp or pointed objects work their way through the wall of the reticulum, requiring surgery to save the goat's life.

3. **Omasum:** Also called many ply, this compartment has many folds, similar to a cabbage. Its function is to absorb nutrients.

4. **Abomasum:** This is the second largest of the four stomachs and is referred to as the "true stomach" because it is where real digestion occurs. The abomasum contains pepsin enzymes and hydrochloric acid that break down proteins into easily digested simple compounds.

When a ruminant eats, the partly chewed food enters the rumen and reticulum, where it ferments with the acidic saliva. A well-developed rumen can hold 4 to 5 gallons of liquid and fermenting fiber. What cannot be broken down is sent back to the mouth as soft masses (cud) to be rechewed before entering the rumen and reticulum again. The walls of the rumen and reticulum absorb fatty acids and vitamins. After the plant matter is thoroughly broken down, it enters the second two chambers, the omasum and abomasum, where nutrients are extracted and final digestion takes place.

The digestive systems of kids (baby goats) do not function in the way those of adult goats do. Newborn kids use only the abomasum (true stomach) for digestion of liquid food. As the kid begins to experiment with eating roughage, its rumen begins to develop, followed by the other chambers. Once you notice a kid chewing its cud, the four stomach chambers are likely fully formed and functional.

In order to maintain the rumen in good working order, a goat requires the right proportion of roughage to grain. Feeding too much grain to a goat is not only expensive, but it impairs the muscle tone of the rumen. The digestive systems of adult goats that are not fed enough fiber begin to function more like those of single-stomached animals, and the contractions of the rumen almost cease. On the other hand, when the goats have too much fiber without enough grain to provide energy for digestion, impaction of the rumen may result. The rumen becomes so full that food cannot pass into the omasum, which results in bloating and death.

Udders

Proper care of a dairy goat's udder is essential to milk production and the health of the doe. The udder of a goat has two halves, each containing a single mammary gland; it has two teats where a cow's has four. The two sides of the udder should be the same size and separated by a cleft. The mammary glands contain tiny sacks of cells called alveoli (or alcilli) that secrete milk, surrounded by muscular cells that allow the milk to exit, also called "letting down" the milk, when the udder is stimulated by milking or a kid sucking.

The udder is suspended from the abdominal wall by a series of ligaments called udder attachments or udder supports. If these become abnormally stretched or weakened for any reason, the udder hangs too low and is prone to injury, such as being stepped on by the doe or by other goats.

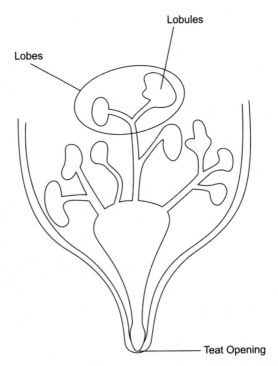

Diagram of a goat's udder

Here is an overview of the visible parts of the udder and their respective functions:

- **Udder cistern:** Also called the gland cistern, this is the part of the udder that appears full, the area above the teat that holds the milk. Each cistern holds about 1 pound of milk at a time.

- **Teat cistern:** The reservoir between the streak canal and the udder cistern, through which the milk passes on its way down and out of the udder.

- **Streak canal:** The final passageway for milk out of the teat.

- **Furstenberg's rosette:** A many-folded mucus membrane that works as a plug to prevent leakage. It also keeps bacteria from entering the streak canal and traveling up into the udder.

These parts of the udder are invisible to the naked eye:

- **Alveoli (plural):** Tiny sack-like structures containing cells that secrete milk when hormonal conditions are appropriate. These cells are surrounded by muscle cells that contract during milk "let-down."

- **Lobules:** Bunches of alveoli bound together.

- **Lobes:** Bunches of lobules.

- **Myoepithelial cells:** These cells surround the alveoli and release a hormone called oxytocin that stimulates the udder to let down the milk.

- **Milk canals:** A series of tubes that carry milk from the alveoli to the udder.

The alveoli secrete milk, which then travels down the milk canals to the udder cistern, where it accumulates. At the time of milking, the myoepithelial cells release oxytocin, which causes the milk to travel from the udder cistern through the teat canal to the streak canal and out into your milk bucket or jug. After milking, the Furstenberg's rosette closes, effectively plugging the entrance to keep out bacteria and prevent leakage.

Milking at regular 12-hour intervals, massaging and washing the udder prior to milking, and using teat dip and an udder balm maintain the health of the udder

Horns

Goats are horned animals, for the most part, though a few breeds are naturally polled, or born without horns. Both the male and female of the species carry horns. The length and size vary with the breed. Certain breeds, such as the Saanen, are purposely bred to be hornless, but this has some consequences. The gene for hornlessness is linked to a gene for the development of hermaphroditism (the presence of both male and female reproductive organs). Female goats that received a hornless gene from both parents (homozygous) will be barren, and bucks may suffer from blocked semen. Horned goats and polled goats that received the hornless gene from only one parent do not have these problems. To avoid producing sterile offspring, a hornless female should always be bred with a horned buck.

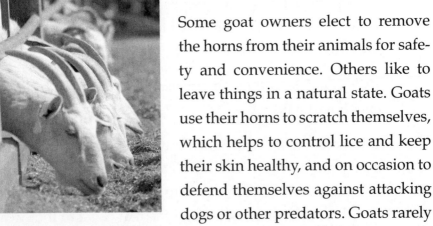

Some goat owners elect to remove the horns from their animals for safety and convenience. Others like to leave things in a natural state. Goats use their horns to scratch themselves, which helps to control lice and keep their skin healthy, and on occasion to defend themselves against attacking dogs or other predators. Goats rarely wound people intentionally with their horns, but they can use them to damage fences, shelters, and equipment. If polled goats and goats with horns are kept together in the same herd, they should be watched carefully. A dominant or aggressive goat may use its horns to injure others.

Dairy goats and goats used for shows routinely have their horns removed because it makes them much easier to handle and removes a possible source of injury. Goats with horns cannot fit their heads easily into milking stands or through feeding racks fitted with "keyholes," slots with an opening at the top to allow the head to enter, to control feeding behavior. They are also more likely to get their heads caught in a fence or a hole in an enclosure or building wall.

Horns are removed with a relatively simple procedure called disbudding when goats are young and the horns are still buds. This is the safest horn removal option. Removing larger horns as the goat ages is a surgical procedure that should only be performed by a vet and is painful and traumatic for the animal. It requires removing skin from the goat's forehead and leaves gaping holes that take time to heal. *(For more information on disbudding, see Chapter 9.)*

Teeth

A goat has front teeth only in its lower jaw; the upper jaw has a hard, tough pad called a dental palate. For good browsing efficiency, the lower incisors should align with the leading edge of the dental palate. If the teeth protrude beyond this edge, the goat has a condition known as "monkey mouth" or "sow mouth," and if the teeth meet the palate behind its forward edge, the condition is called "parrot mouth."

When a kid is born, it often has its first two front milk teeth and sometimes the first four have already broken through. By its third month, all the milk teeth have broken through. At about 15 months, the adult teeth begin to appear, and by the time the goat is 3 ½ to 4 years old, it has all its adult teeth. An adult goat has 12 molars in the upper and lower jaws and 8 front teeth in the lower jaw. A goat with all its adult teeth is "aged." When the goat is 5 years old, the front incisors begin to spread apart and wear down, and eventually they break off and fall out. A goat with missing front teeth is "broken-mouthed." When all the teeth are gone, the goat is a "gummer." It is possible to tell the age of a young goat by looking at its teeth, but once it is more than 5 years old, the wear and condition of the teeth may vary according to its diet and the location in which it lives. A coarse diet wears teeth down faster than a soft diet.

Development of Goat Teeth

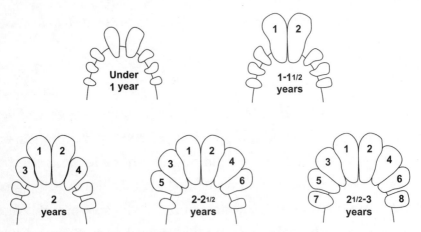

A mature goat has eight teeth in its lower jaw.

Safety and Goats

Goats are friendly and generally nonaggressive; however, they can knock you down. Also, if they are horned, you can accidentally injure yourself simply by turning too quickly and running into a goat. If you are small in stature, or if you have children who plan to help with the goats, you must keep safety in mind when selecting animals for your herd. Dairy goat breeds vary greatly in size; some goats weigh 150 pounds. The temperament of an individual goat is also a consideration; a particularly rambunctious or malicious older goat may be too much for a child to handle.

The more familiar you are with an animal, the safer you are. Many barnyard injuries occur because someone does not know how the livestock will move or how they will react. The more you understand your goats, the more confidence you will have in handling them, and the more pleasant your experience will be. Goats that have been bottle-fed as kids or handled frequently as they were growing up are easier to manage and milk because they are used to close contact with humans. When you purchase goats in a private sale, rather than at an auction, you will be able to see how their previous owner treated them and to learn something about the nature of each goat.

Goats are normally friendly and enjoy the company of humans, but a goat that is injured, frightened, or nervous might not be as friendly. A buck that is eager to mate will be more assertive than usual. Males should be handled and trained while they are small; if they are not, they grow into big teenage goats that want to play with you and may deliver a powerful head-butt. A goat will rear up high on its hind legs before butting anything, so you have a few moments' warning. A goat that does not want to be milked might use its horns or its bare head to toss a pail across the room, toss a passing cat, or kick out at you while in the milking enclosure. These occurrences are uncommon, but keep them in mind, and be aware of what could happen.

> ## TIP: Do not treat a kid like a puppy
>
> When kids are small and cute, it is tempting for children to play with them like puppies, pushing against their heads. This can inadvertently teach them to butt people when they become adults. Play with kids by running with them, chasing them, petting and scratching them, teaching them to walk on a leash, and to obey commands or hand signals.

Dairy Goat Breeds

The American Dairy Goat Association (ADGA) lists eight major goat dairy breeds based upon average annual milk yield: Alpine, Saanen, Sable, Toggenberg, Oberhasli, LaMancha, Nubian, and Nigerian Dwarf. The first five on the list are often referred to as "the Swiss breeds," and they all originated in the mountains of Switzerland and eastern France. In addition to these eight, various other breeds developed in the recent past or are not as well known are also used in dairy production.

Chapter 4 explains the difference between purebred goats and goats of mixed breed and how to select the goats for your dairy herd.

Alpine

Sometimes called French Alpines, these goats are descended from a herd of 21 goats that arrived in the United States from France in 1920. They are best known for being highly adaptable and will thrive in any sort of climate. Alpines are also hardy and generally enjoy good health, even in difficult environmental conditions.

Alpines have one of the highest average annual milk productions of all the dairy breeds: more than 2,000 pounds annually with 3.5 percent butterfat.

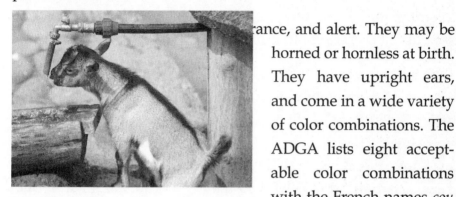

ance, and alert. They may be horned or hornless at birth. They have upright ears, and come in a wide variety of color combinations. The ADGA lists eight acceptable color combinations with the French names *cou blanc*, or (white neck;), *cou clair*, (clear neck;), *cou noir*, (black neck;), *sundgau*, (black with white markings;), *chamoisee*, (brown or bay with black head, feet, and stripe down the back;), *broken chamoisee*, (chamoisee with a splash or band of another color;), *two-tone chamoisee*, (light front quarters with brown or gray hindquarters;), and *pied*, (spotted or mottled). The coloring is often very interesting and dramatic, including tan, red, bay, or brown backgrounds with white, black, and gray lines, spots, and stripes in various patterns.

The Alpine is a medium to large-sized goat. The doe stands about 30 inches tall at the withers (shoulder blade) and weighs an average 135 pounds. The buck stands 32 inches at the withers and weighs about 160 pounds. However, Alpines can grow to weigh as much as 250 pounds.

Saanen

The Saanen (SAH-nen) is the breed most people think of when they think of a dairy goat. It was made famous in the children's book *Heidi* — a big, white, friendly goat with flowing beard that happily wanders over the mountainsides. Saanens are sometimes called the Holsteins of the dairy goat breeds because of their high average milk output. They are well known for their calm, easy-going natures, their ability to adapt to changes in the herd, and their tolerance for a variable climate. Because white goats are prone to sunburn, it is essential to provide them with lots of shade.

Saanens come from the Saanen Valleys in Switzerland and then were spread throughout Europe beginning in 1893. Saanens first arrived in the United States in 1904. They are the largest of the dairy breeds. It is thought that there are more Saanens in the world than any other breed of dairy goat. Depending upon the year, Saanens or Alpines generally yield the highest annual milk production. Currently the yield of the Saanen is much more than 2,300

pounds per year with butterfat content of 3.5 to 4 percent.

Saanens are white to cream in color and have erect, forward-leaning ears. Saanens may have horns or be naturally polled. They carry themselves with a dignified air. The doe should be at least 30 inches tall and weigh at least 135 pounds; the buck should stand at least 32 inches at the withers and weigh at least 160 pounds. It is not common for Saanens to grow well beyond these averages to reach heights of 35 inches or more.

When shopping for Saanens, watch for unshapely udders and misshapen backs, which leave weakened hindquarters. Both of these faults occur in the breed, but if you find goats that do not show such faults, you will have dairy goats that are unusually sturdy and strong, vigorous, and rugged of bone.

Sable

The Sable is often called "a Saanen in party clothes" because it is actually a Saanen that is not purely white. In recent years, these goats attracted a loyal following that lobbied the ADGA to grant it status as a separate breed, which it recently did. The Sable has the same traits as the Saanen in terms of size, milk output, and temperament. Like their Saanen cousins, these goats are heavy milk producers and are willing to milk through an extended lactation when encouraged to do so.

Toggenburg

The Toggenburg is not only considered the oldest goat breed to have an official registry. The first herd book, containing the list and pedigrees of one or more herds, was established in the Toggen-burg Valley of Switzerland in the 1600s. This goat is another of the strong, sturdy, adaptable Swiss breeds. Toggenburgs were the first purebred goats to arrive in the United States, in 1893. Today, they are the fewest in registered number for major dairy breeds.

The Toggenburg does better in a cooler climate. It is only slightly behind the Alpine and Saanen in milk production, with a butterfat content that is currently at about 3.7 percent. According to the Dairy Herd Information Association (DHIA), the doe that holds the all-time record for milk production was a Toggenburg that gave 7,965 pounds over a 305-day period. Many Toggenburgs milk well into their teen years.

Toggenburgs are various shades of brown — from fawn to bay to chocolate to espresso — with dramatic white markings on the face, the rump, and the legs. Like the Saanen, their ears are erect and forward pointing. The Toggenburg often has a wattle and a pronounced beard. Toggenburgs are a medium-sized breed, with does weighing at least 115 pounds and commonly weighing as much as 150. Toggenburgs bred in Switzerland and Great Britain tend to be much smaller and have longer hair.

The Toggenburg is often used as a show goat by children in 4-H clubs or adults starting out in the show world because it has a nice, calm disposition and adapts well to travel and new environments.

Oberhasli

In 1978, the official registry for the Swiss Alpine changed the breed's name to Oberhasli (oh-ber–HAHS-lee). The Oberhasli was developed near Bern, Switzerland, and shares traits such as sturdiness and adaptability with other Swiss breeds. It is less common in the United States than some of the other breeds, and at one time was listed as endangered; however, it is recently gaining favor for show and dairy purposes because of its smaller size, attractive appearance, and easy temperament.

The Oberhasli is known for its unusually sweet temperament. It is interesting, though, that when integrated into a mixed herd, the Oberhasli doe will often become the queen. These goats have an alert and inquisitive expression that endears them to their owners. Like many goats of the Swiss breeds, they tend to be vocal.

The Oberhasli is also a medium-sized dairy goat. The doe is at least 28 inches high at the withers and weighs at least 120 pounds; the buck is at least 30 inches high and weighs 150 pounds. The Oberhasli may be either horned or polled at birth. The males develop a thick beard as they mature, but the does have a few scraggly hairs or no beard at all. They generally have wattles at birth. The coat is a rich, deep reddish-brown color (formally called *chamois*) with black stripes on the face, back, udder, and belly. An occasional doe will be black due to a recessive gene in the breed; this color is not suitable for a show animal.

Oberhasli milk production annually averages more than 1,600 pounds, with about 3.6 percent butterfat. The milk of this breed is distinguished by a sweet taste.

LaMancha

The LaMancha is a dual-purpose milk and meat breed. Its ancestors were introduced into the American West by Spanish conquistadors who arrived in California from Europe. Their goats were believed to have originated in Spain, though this is not certain. In the early 20th century, these goats were bred with various Swiss breeds to arrive at a new breed. In 1958, after some hard lobbying by aficionados, the first LaMancha was registered by the ADGA. The most distinguishing feature of the LaMancha is its ears, which are either extremely short, called an elf ear (about 2 inches), or are almost nonexistent, called a gopher ear. The ears have little or no cartilage and usually curl at the tips.

This breed is known to have an uncommonly calm, gentle, but curious personality. Because they are one of the medium-sized breeds, LaManchas make good show ring goats for children involved in 4-H Club.

The LaMancha can withstand a great deal of hardship in its environment and still yield high milk amounts, with excellent butterfat. On average, LaManchas will give an annual yield of about 1,700 pounds of milk, with a high average butterfat content of 3.9 percent.

LaManchas are known for their longevity and for many years of producing high quality milk. The does stand about 26 inches high at the withers and weigh about 130 pounds on average, but they can be bigger. Males weigh about 150 pounds. They come in a variety of colors and color patterns.

Nubian

Nubians are known for their long, pendulous ears. Many people favor them simply for their looks. Adding to their appeal is a short, glossy coat of fine hair.

Currently, Nubians are the most popular dairy goat in the United States. The breed was originally used as a dual-purpose meat and dairy goat. In the United States, the Nubian is generally thought of solely as a dairy goat, but if you want to raise a dual-purpose breed for profit, you might want to consider it.

The history of the Nubian is unique and complex. For several centuries, English merchant ships sailed the world with English goats on board to supply food for their crews. Once in foreign ports, these goats bred with native goats. Over the years, a distinct type developed that was a mix of the original English goat with the traditional long-eared breeds of India and Africa. During the 19th century, many of these mixed goats were brought from Nubia, in North Africa, through France, finally arriving about 1883 in England. The Nubian was registered as a distinct breed in England in 1896. In the following decades, the instantly popular goats were imported to the United States, and by 1920, more than 40 individual Nubians were registered.

Although its average milk yield is lower, the Nubian has a few advantages over the Swiss breeds. For one, its milk is high in butterfat, which makes it an excellent choice for cheese production. The Nubian is thought of as the "Jersey [cow] of the dairy goats," because there is less milk, but the butterfat percentage is extremely high. The average annual yield of the Nubian is about 1,600 pounds annually, with butterfat content averaging 4 to 6 percent. It is not uncommon for individual goats to exceed the 6 percent. The Nubian has a longer breeding season than the Swiss breeds, and so can be milked off-season for continuous production.

Nubians come in any color and some are spotted. The doe stands about 30 inches tall at the withers and weighs an average 135 pounds. The buck stands 35 inches high and weighs up to 175 pounds. Nubians are built a bit differently from the Swiss; the backline is not straight but rises toward the rump, and the back often has a bit of a dip. The back legs, though sturdy, are long.

Nubians are not only vocal, but are unusually loud, which may be a factor if you have neighbors living in close proximity to the herd. Be aware: The Nubian is a jumper, so high, sturdy fences are necessary to keep this breed contained.

Nigerian Dwarf

Early in the 20th century, zoos imported small goats for feeding their large cats. Over time, it was discovered these little goats made affectionate, fun, friendly pets. By the later part of the century, these dwarf goats — so called because their heads, bodies, and legs are proportionate — were divided into two breeds, the Pygmy and the Nigerian Dwarf. The Nigerian Dwarf has proven itself in the dairy world with high milk production and butterfat and protein percentages comparable to the large breeds. For this reason, the ADGA designated it an official dairy breed in 2005.

In addition to its physically manageable size, the goat breed offers various advantages, such as being a dual-purpose milk and

meat goat, and it breeds year-round. This means you can stagger breedings so there are lactating goats consistently available to milk any time of the year. Because it is small, the Nigerian Dwarf requires less space and eats less. If you are looking for a goat to learn with, or simply to supply your household with milk, the Nigerian does not give a huge quantity of milk per day.

There are benefits to using Nigerian Dwarfs for a dairy. Nigerians are generally slender, with teats comparable in length to those of a larger goat, so they are not difficult to milk. If you acquire Nigerians for milking, be careful to purchase goats that are accustomed to it. If you purchase a goat that was previously a pet and not used to milking, you may be in for a hard time. When used as a dairy goat, the Nigerian Dwarf gives an average 2.8 pounds a milk per day, but can give as much as 8 pounds, and over a year can produce up to 800 pounds or more. Butterfat content is high, at about 5 percent and rising to as much as 10 percent toward the end of the lactation period. This milk is used particularly for making cheese or soap.

The Nigerian Dwarf is a favorite for 4-H Club projects because of its small size. This goat particularly enjoys the company of humans and can become quite attached (and vice versa). Individuals are often "a big goat in a small package" and exhibit a lot of personality.

Nigerian Dwarfs come in a variety of colors; have upright, erect ears; and have fine hair, ranging in length from short to slightly longer than short. Some Nigerians have blue eyes, and most have horns at birth. Because height is a factor in classification (or registration) as a dwarf, it is an important consideration. In general, the Nigerian Dwarf is going to be about half to two-thirds as big as the smaller large breeds. There are actually two different height standards for this breed. Goats registered with the ADGA should stand no higher than 22.5 inches at the withers for the doe, and 23.5 inches for the buck. However, if the owner would like to register the goat with the Nigerian Dwarf Goat Association (NDGA), the height for a doe should be between 17 and 19 inches, and the buck should be no taller than 19 to 21 inches; for both genders, the ideal height never exceeds 21 inches.

The four or five other breeds

The African Pygmy (or just Pygmy) is similar to the Nigerian Dwarf. It is a dual-purpose goat frequently used as a pet. It has gained popularity as a dairy breed; however, it is not registered as such with the ADGA but is registered with the National Pygmy Goat Association. This goat is only 16 to 23 inches tall at the withers and the doe averages a weight of 55 pounds. The Pygmy can produce as much as 4 pounds of milk a day and up to 700 pounds a year. The lactation period is shorter than that of a full-size breed — four to six months as opposed to ten months — but the butterfat content often exceeds 6 percent, making this goat an excellent candidate for cheese- and soap-making enterprises.

Individuals can be somewhat difficult to milk because of their stocky, low bodies. If you plan to milk your Pygmy, make certain before purchase that the goat is accustomed to being milked, that you are physically able to get under the goat to milk, and that the teat length is going to be adequate for your hands; you should be able to fit your thumb and two fingers on the teat. Pygmies tend to have triplets or even quadruplets, so you could have your hands quite full with kidding.

Recently, several mini-breeds have developed from the Swiss breeds and LaMancha. Each of these shares traits with its parent breed. Some newer breeds are currently being developed; one breed with an enthusiastic and growing following is the Kinder. This breed was developed by crossing pygmies and Nubians and is a good producer while being a bit smaller than the large dairy goats. This is a dual-purpose breed with excellent milking ability. The Kinder gives an annual milk yield of up to 1,500 pounds, and its milk has a butterfat content of 5 to 7 percent. Originating in 1985 at Zederkamm Farm in Washington state, this breed now features more than 50 herd names and hundreds of registrations with the Kinder Goat Breeders Association. With a growing fan base nationwide, it is a new breed that is easy to find and purchase, and it is only a matter of time before it will be registered with the AGBA.

Chapter 3

Planning for Your Dairy Goats

ow that you have decided you want to try your hand at raising dairy goats, it is time to make physical preparations for their arrival. The more prepared you are on the day you bring your goats home, the more pleasant the start-up process will be. Your new goats will be under considerable stress as they adapt to living in a new environment with unfamiliar caretakers, and you will be challenged as you work to establish a daily routine and learn how to manage and care for them. It will be easier for you and the goats if everything is ready before they arrive, and you have all your basic supplies on hand. You do not want to keep your goats barricaded in a shed while you frantically try to finish construction of an escape-proof enclosure or bleating with hunger while you try to locate and purchase ingredients for their feed. It will be best for the goats if their feeding and sleeping arrangements are already in place so they can quickly accustom themselves to their new surroundings. In the beginning, you will need to spend time getting to know your goats and working out a routine for ordinary chores such as feeding and milking, without the added worry of trying to make do without urgently needed supplies and equipment.

The first step is to take inventory of the property where you intend to locate your goats. Goats do not require as much space as larger livestock, and you may be able to accommodate them in an existing shed, garage, chicken coop, or barn. In temperate climates, goats can be housed in a three-sided, south-facing open shelter. A shelter with its open side to the south will be sheltered from northerly winds and receive the maximum

Birthing pens at Soggy Bottom Farms.
Photo courtesy of Soggy Bottom Farms.

amount of sunlight and warmth during the day. For dairy goats, you need a draft-free permanent shelter where they can give birth and take care of young kids and an enclosed area for milking. Whatever arrangements you make, there should be enough space to move around with ease when milking and cleaning up. Running water and electricity make your work easier but are not absolute necessities. Your goats will need to be contained with a sturdy, escape-proof enclosure. Goats will need shade, but any trees within reach of your goats will quickly be stripped of their bark, lower branches, and leaves, and their trunks must be protected with sheaths or sturdy wooden structures.

The next step is to create a budget. When you are first considering goat farming, you may not be aware of all the extra costs. To avoid unpleasant surprises, make a complete list of your expenses, which will include the following:

- Cost of the goats
- Transporting the goats and their products
- Veterinarian feeds
- Milking equipment
- Fencing and building supplies
- Lighting
- Feed and bedding
- Medical supplies

Determine how much feed and bedding you will have to buy and how much is naturally available on your property. Although goats are expert foragers, they need grain and dietary supplements, especially when lactating and breeding. The type of feed a goat receives affects not only its health and weight, but also its reproductive capability and even the taste of its milk. The feed you have to purchase from month to month will depend upon the quality and type of vegetation on your property, whether you are breeding goats year-round, the local climate, and the number of goats you are caring for.

After feed, your largest expense will be veterinary care. Inevitably, you will need a veterinarian's services at least a few times a year. Even if you are an advocate of the current anti-vaccine movement, some vaccines are absolutely necessary, and some may be required by law (check with your veterinarian or local agricultural extension office). Failure to protect your herd from disease will not only hurt you economically, but it will also threaten the health of your neighbors' livestock and other animals on your own farm. You will be purchasing worming medications and administering them at regular intervals during the year; the type and frequency will vary according to your climate and location. Because your goats are dairy goats, your enterprise will involve breeding, and during pregnancy and birthing, occasional emergencies require a vet's assistance.

Visit a local veterinary supply store to price vaccines and worming medicines. Calculate your costs considering the frequency of use recommended for each product, which will vary from three-month intervals to once yearly. Your local agricultural extension office and other goat owners can help you estimate your expenses.

Researching and developing a working budget will help you navigate your first year as you learn the ropes. Anticipating the cost of keeping your goats will prevent unpleasant surprises when unexpected expenses arise.

It will be some time before you can realize any income from your goats, but you may be able to save money by buying used equipment on Craigslist (**www.craigslist.org**), or through classified ads, and by using materials that you already have on hand. You will also need to locate local suppliers of feed, grain, and other items you will need to buy regularly. Other costs to consider are your water bill and the electrical bill for heating or cooling barns, the use of mechanized dairy equipment, and other tools. *Chapter 11 explains how to assess the start-up expenses for a commercial dairy business.*

Finally, you must plan how you are going to care for the goats. Someone must be available to milk the goats at regular 12-hour intervals, and you will have to work this into your daily routine. To keep your herd healthy and productive, feeding, cleaning, and basic care must be done on schedule. If this is a family project, it is important that family members are willing to accept some of the responsibility for chores. If they do not want to participate initially, do not assume they will change their minds later on. You must be prepared to do the work yourself or hire someone to help.

Containing and Managing Your Herd

Dairy goats have more requirements than goats raised for meat or fiber or kept as pets. In addition to shelter and feeding areas, you will need space for milking, for processing milk and storing equipment, and for taking care of kids and pregnant does. Everything has to be kept as clean as possible to protect the purity of the milk and the health of the does.

Milk production and quality is closely linked to stress levels in your goats, so you want to make their environment as pleasant

as possible by providing the most comfortable accommodations you can afford. Goats need places where they can stay warm (in cold weather) or out of the sun (in hot weather) and sleep safely; pens or enclosures where they can be separated from other goats from time to time; an outdoor run or exercise yard; pastures or fields where they can forage; and objects to climb on. Feeding and watering stations should be easily accessible and easy to keep clean. You will also want to provide stall areas that can be easily cleaned and pens that are safely separated from storage areas.

How many pens and pastures, and how much space you need, depends mostly upon the nature of your region's climate, the goals you have in mind for your operation, and of course, the size of your herd.

The basic requirements for dairy goat housing are:

- Space
- Ventilation
- Drainage
- Water
- Flooring
- Bedding
- Feeding area
- Exercise area
- Milking parlor and milking stand
- Storage for feed, grain, bedding, milking supplies

If you are starting with just two or three goats, keep the arrangements simple. You can expand your facilities later on when you decide to increase your herd. Before constructing any special buildings or investing in expensive dairy equipment, consult your local agricultural officials and zoning department to make sure any improvements you make comply with zoning laws, building codes, and regulations. Local agricultural bureaus will have information and advice specific to goat keeping in your region, including recommendations for housing, building plans, and warnings about disease.

To determine how much land you need for your goats, first check local government zoning regulations. Besides being zoned for livestock, you might have to comply with regulations specifying the amount of land required per animal housed. A minimum amount of space is probably required to house any livestock at all. In most rural areas, the rules governing goat keeping will not be too restrictive for your purposes. Some urban areas do not allow livestock to be kept at all while others are friendly to community gardens and small farm enterprises. In some areas, goats are defined as livestock; while in others, they are defined as companion animals. This definition can affect the success of a dairy farming operation.

Most people are surprised to learn that, when well managed, five acres can support 100 dairy goats. You can keep a dozen goats on one or two acres, including land for your stable. Within this space, specific areas and enclosures should be allotted for various purposes.

Before you begin arranging your own facilities, visit as many local goat owners as possible and observe how they manage and house their goats. You will pick up some good ideas, and you can learn how they have handled specific difficulties.

Fences and Enclosures

No animal can escape like a goat. Goats are clever and like to work at solving problems. A small flaw in a fence presents an irresistible challenge to a goat brain. Goats crawl, twist and turn, leap, and even climb to get to the other side of an obstacle, greatly enjoying themselves in the process. Your mission is to eliminate the opportunity for this goat play. Prepare an escape-proof enclosure before you bring home your first goats. Goats love to eat shrubs, vegetables, garden plants, and tender rose bushes (which contain a lot of vitamin C). They also love to jump on top of cars. A reliable, well-made fence is essential to everyone's well-being and peace of mind.

A well-built fence remains strong for as long as it is needed, prevents goats from going where they do not belong, and protects them from dogs, coyotes, and other predators. Good fences also protect goats from demolishing your trees and shrubs. For keeping goats, you will need three types of fencing: boundary or perimeter fences, interior fences, and barn lot fences.

- Fences around the perimeters of your pastures are the longest fences and are usually permanent; they should be constructed of quality materials that do not require much maintenance.

- The interior fences used to divide pastures up into sections for rotation, or to separate goats, can be either permanent or temporary and movable. These do not need to be as durable and strong because even if the goats escape, the perimeter fences will still keep them in the enclosure.

- The fences that enclose your barn lot or goat yard must be able to withstand a lot of wear and tear and abuse from the goats.

TIP: Goats like to put their feet up

Whatever facility you are constructing for goats, remember that they like to stand on their back legs and rest their front feet on any available object or projection. This means your wire fences will eventually begin to sag, dirt will get into watering pans, and hay will be dragged down from feeders. Horizontal elements of wooden fences and gates should be on the outside of the goat enclosure, not inside where the goats can climb on them.

A small, fenced-in exercise yard adjacent to your goat enclosure is ideal, even if you have a large pasture for your goats. This should be in a sunny, dry spot, preferably on the southern side where it will get the most exposure to sunlight. A slight slope will help to keep it drained and dry. If the yard tends to be muddy or damp, at least part of it should be paved with concrete or pavers for easy cleaning. Walking on this hard surface will help to keep hooves trimmed. The fencing around this enclosure must be particularly goat-resistant because it will take more of a beating; confined goats will entertain themselves by attempting to destroy the fence.

Several types of fencing are appropriate for goats, including wire fencing, wooden, and electric fences. Your selection should be based on your budget and how you intend to manage your goats.

Woven wire fencing

Wire fencing is widely used by goat keepers. It consists of smooth, vertical wires held in place by horizontal wires called stays. The vertical wires are spaced 6 to 12 inches apart. The horizontal wires are generally closer together at the bottom and wider apart at the top. Wire fencing is available in galvanized, high-tensile (able to resist stretching), and polymer-coated, high-tensile varieties. The

numbers on the packaging tell you the size and spacing of the fence; for example, 8/32/9 fencing is 32 inches tall, has eight horizontal wires, and vertical wires every 9 inches.

Photo courtesy of Soggy Bottom Farms.

Typically, wire fencing 4 to 5 feet high is used and is attached to 7-foot posts spaced 12 feet apart. A strand of electric wire 12 inches from the ground and another strand at about the shoulder-level of an adult goat should discourage the goats from leaning on the fence or rubbing against it. Many goat keepers use wire fence 32 inches high and augment the height with several strands of electrified or barbed wire above it.

The disadvantages of wire fencing are its cost and the time and expertise needed to install it properly. Once installed, a woven wire fence is more or less permanent. It is durable but needs to be checked regularly and repaired if it starts to bend or sag. Horned goats enjoy tugging at the wire with their horns and sometimes get their heads stuck in the wire squares. Using wire with 12-inch spacing between the vertical wires can mitigate this problem.

Strong wooden fence posts must be used for corners and gates, but in between them, you can use T-posts, which are metal posts, often made from recycled railroad tracks, that are pounded into the ground with a handheld post pounder.

TIP: Choose fence posts carefully

Fence posts are treated to make them durable and resistant to rot and insects. Because dairy goats might gnaw on them, pay careful attention to the chemicals used to treat the wood. Pentachlorophenol is effective in preventing decay and insect damage, but contains carcinogenic compounds called dioxins. Posts treated with creosote may damage the hides of sensitive livestock.

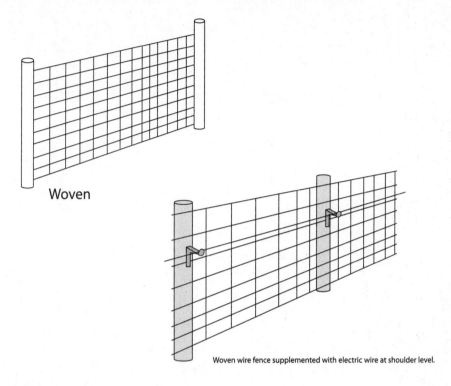

Woven

Woven wire fence supplemented with electric wire at shoulder level.

High-tensile wire fencing

High-tensile fencing, developed in New Zealand, uses several strands of smooth wires held along posts, or combinations of posts and spacers called battens, stays, or droppers. This fencing system uses smooth 12 ½-gauge wire with a yield strength of 1,600 pounds. (A conventional 12 ½-gauge wire will yield at tensile force of less than 500 pounds and break at less than 550 pounds.) Each wire is stretched with 250 pounds of tension, maintained by permanent in-line stretchers or tension springs. The higher tension in the wire reduces sagging but requires the use of strong end- and corner-brace assemblies. High-tensile fencing is less expensive than barbed or woven wire fences, requires less time to erect and repair, requires fewer posts, is easier to handle, lasts longer, and is easily electrified.

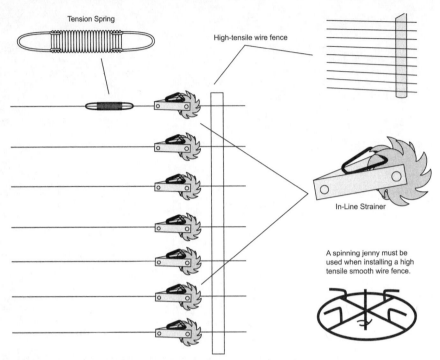

Example of a high tensile wire fence

TIP: Non-electrified steel wire fences must be grounded

All non-electric fences using steel wires on nonconductive posts must be grounded to protect humans and livestock from electric shock due to lightning and fallen electrical wires. Grounding electrodes should be a standard galvanized steel post or a ¾-inch galvanized steel pipe driven in firm earth to a minimum depth of 3 feet. A ground should be installed every 300 feet in moist or damp soils. Grounds should be used every 150 feet in dry, sandy, or rocky soils. All fence wires should be connected to the ground rod. Wire fences should be broken up at maximum intervals of 1,000 feet by means of a wooden gate, wooden panel section, or insulating material.

Wooden fences

A fence made of vertical wooden palings 47 inches high, with the bottom no more than 2 inches off the ground and no more than 2 inches of space between the palings, is largely goat-proof. Goats can squeeze through small spaces, so the fence should be regularly inspected to ensure all the palings are sound and firmly attached. A wooden fence can be reinforced with wire fencing to make it escape-proof. Wooden fencing is usually too expensive to be used for large enclosures.

Electric fences

Some goat keepers swear by electric fencing, but goats can occasionally escape through it. It is better used to reinforce a wire or wooden fence, or for interior partitions within a larger enclosure or pasture. Electric wire of various types and sizes is available at farm stores, as well as electrified string net, polywire, polytape, and rope fences. These fences are easy to move and rearrange and are ideal for temporary fencing, such as when you are moving your goats around different areas of a pasture. They are designed to work with step-in fence posts, made of metal, plastic, or fiberglass that can be inserted into the ground using your body weight. Electric fence chargers cost between $60 and $300, depending on the length and sophistication of your fence system. Fi-Shock™ (**www.fishock.com**) and Parmak® (**www.parmakusa. com/index.html**) sell solar-powered electric fence chargers that are convenient for remote locations.

Goats can be trained to fear an electric fence by luring them toward it with grain until they experience the electric shock.

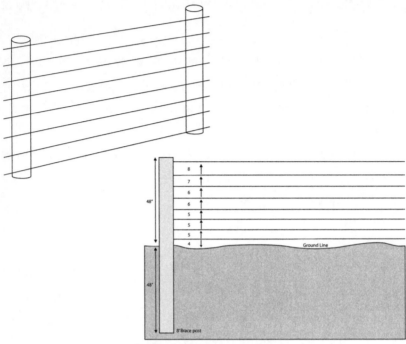

Example of an electric fence

Barbed wire

Barbed wire can be used to goat-proof existing farm fences, but it is not generally recommended for dairy goats because it can injure their coats and udders. To contain goats, a barbed wire fence should be built of eight to ten strands of tight, evenly spaced 15-gauge wire.

Livestock panels

Livestock panels made of welded steel rods are excellent for temporary enclosures such as paddocks and pens, but like wooden panels, they are usually too expensive for fencing. They are easy to install and move around. Sharp edges and snipped wires should be smoothed with a file to avoid torn clothes and injury to your goats. Sheep stock panels have smaller spaces between the welded wires so small animals will not get their heads caught, but these are even more expensive.

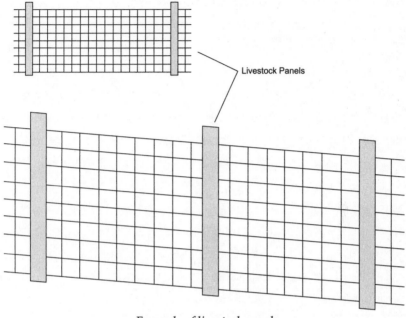

Example of livestock panels

Gates

Fences, as well as gates, must be made escape-proof. Gates should open inward so goats cannot push their way out. Some goat keepers install strong springs that snap the gates shut. Place latches low on the outside of the gate where goats cannot lean over and reach them. Crossties and frames should be on the outside of a gate where the goats cannot use them to get a foothold. Gates and fences should never be topped with sharp pickets because the goats can get their heads and hooves stuck between them.

Entertainment

Goats love to jump onto raised objects. You can provide entertainment by building an exercise shelf or placing old tractor tires, wooden crates, barrels, and other items in their enclosure for them to play with. Do not place these items in close proximity to a fence where they could help a goat escape.

TIP: Protection for your trees

Goats will soon strip the bark off the trees in their yard or pasture. To keep fruit or shade trees alive and healthy, you must protect their trunks. For young saplings, you can purchase 5-foot metal mesh or plastic tubes, available in garden stores that are designed to wrap around trees and protect them from deer. The trunks of larger trees can be wrapped with rabbit wire, mesh, plastic gutter covers wired together, or layers of plastic screening. Inspect the trees from time to time to make sure the wrapping is not inhibiting the tree's growth. If you have only a few trees to protect, you can build a triangular wooden structure around them with sturdy wooden lathes.

Sheds, Stables, and Barns

You can keep goats in sheds, old chicken coops, barns, and garages made of wood, stone, concrete, cement block, corrugated iron, and even cement half-pipes. The type of shelter you use will depend upon several variables, the first of which will be your climate. Severe winter weather is not a concern in a mild climate, but you will need to keep your goats from overheating during hot summer months. Some goats are sensitive to heat, particularly if humidity is high. If your farm experiences snow, harsh wind, and frozen ground for several months out of the year, your shelter will need insulation and protection from the cold.

Dairy goats should be kept happy and comfortable because stress can harm milk production. If your goats are fortunate enough to have a large pasture area, a three-sided field shelter should be located in the part of the pasture farthest from the barn so they can easily find refuge from hot sun, rain, and cold wind. In a cold climate, a low shelter no more than 5 feet high in the front and 3 to 4 feet high in the back will hold the goats' body heat in around their bodies.

In addition to field shelters, you will need a permanent building for long-term shelter, storage, possibly office space, and a milking area. A barn is ideal, but this could also be a smaller structure, such as a stable or garage. If you are able to store feed, bedding, and medical equipment in a separate building away from the area where goats are, you may be able to manage with just a large shed for milking and permanent shelter.

The recommended size for a goat shelter ranges from 12 square feet per animal in a situation where the goats will be outdoors much of the year and have a lot of space for outdoor exercise to 24 square feet per animal in a climate with long, severe winters or hot and humid summers. The more time the goats must spend inside, the more space you will need to allot per animal to keep them comfortable and healthy. If you are planning to breed your does, you need extra space for birthing pens and housing kids. In addition to the living space where goats will spend most of their time, you will need to add extra square footage for storage and milking areas.

Housing Requirements for Dairy Goats

	Large and medium goats	Pygmy goats
DOES		
Open housing	16 sq. ft.	10 sq. ft.
If goats are to be kept indoors year-round	21.5 - 27 sq. ft.	18 - 25 sq. ft.
Yard	200 sq. ft.	130 sq. ft.
Minimum exercise area	50 sq. ft.	40 sq. ft.
Stalls	6X6 ft.	5X5 ft.
Stall partition height	3.5 ft.	3 ft.
Milk parlor	5X8 ft.	5X8 ft.
Fence height	5 ft.	4 ft.
Feeding shelf	15.5 inches wide	15.5 inches wide
If kids are being raised, space per kid	5.5 sq. ft.	
Water	2 to 10.5 quarts daily	2 to 10 quarts daily
Storage – bedding for one year	4 cubic yds.	
BUCKS		
Housing	40 sq. ft.	30 sq. ft.
Stall partition	5 ft.	4 ft.
Yard	100 sq. ft.	70 sq. ft.
Fence height	6 ft.	5 ft.

Your permanent shelter should keep out moisture, wind, and drafts, but at the same time, it needs to be airy and well ventilated, with adequate air circulation. Clean, well-circulated air is a priority because goats are particularly prone to pneumonia, asthma, and other respiratory illnesses, some of which can prove fatal. Condensation on the inside walls of the shelter and an increasingly intense and pungent smell of ammonia are signs that air circulation is poor and more ventilation is needed.

Loose, or open, stables are stables in which the goats are housed together in a central open area instead of in individual stalls. This type of stabling is ideal for goats because they are social animals and like to be together. It also allows you more freedom in managing your goats; you can add an extra goat or two without having to build additional stalls. If you build individual stalls for your goats, use wire mesh, welded pipes, or wooden palings with spaces between them so the goats can see each other and interact. You will need a place where you can isolate a domineering or sick goat, and places where young kids can find refuge from adult goats, such as inside dog crates or barrels.

In addition to a resting area, the goats need a feeding area where they can line up to eat. It must be long enough to accommodate all the goats at the same time. The feeding area should be separated from the resting area by a sill, such as a wooden beam, or be a raised ledge. The area where the goats stand when they are eating should be only lightly littered with straw and swept out daily.

As you assemble your shelter, keep in mind that you will be wheeling a wheelbarrow down aisles and in and out of stalls anytime you transport bedding, refuse, or feed, and leave yourself enough space to maneuver comfortably. Access to electricity and fresh running water will make your work much easier. Without running water, you will have to carry water to your goats twice a day. You will need a sink area for preparing feed and washing equipment and feed utensils.

All gates and doors, such as swinging gates into stalls and gates to the outside, should be hung so they swing inward. You will appreciate why the first time you have a goat trying to crowd you and escape the stall or door. A gate opening toward the inside of an enclosure offers much more control over what goes through that gate when you open it. Goats are notorious for opening latches on doors and gates; make certain yours are durable and lock firmly into place so your goats cannot escape the enclosure or let themselves into the room where hay or feed is stored. It is a good idea to put a goat-proof latch on any refrigerator door or any cupboards that hold medicines.

Temperature

The climate in which you live will determine how much you need to control the temperature and humidity inside your barn or stable. Goats are healthiest and happiest when the air temperature is between 46 degrees and 70 degrees Fahrenheit (between 8 degrees and 21 degrees Celsius). The humidity should be between 60 and 82 percent. Goats are fine in temperatures down to zero F (minus 18 C) as the herd produces its own heat. When the temperature drops below that, you will need to provide some kind of artificial heating. Goats begin to suffer when the temperature rises above 80 F (27 C) and need access to shade and cooling breezes.

Artificial heating and cooling should only be used when necessary. If you live in a climate where you can sustain comfortable temperatures inside the goats' permanent shelter area without artificial heat or air conditioning, you should still have some large ventilation fans hung or installed up high, away from the goats (any electrical cords must be secured away from the herd's reach as well). These will help with air circulation in warm weather, keep the urine smell down, keep stalls dryer, and maintain indoor air quality at an acceptable level for goats and humans.

If you live in a climate where you can maintain a good barn temperature year-round without artificial heating and cooling, you probably do not need to insulate the goat shelter. Insulation helps to keep the shelter comfortable in extreme temperatures. It should be installed behind a sturdy wall away from your goats, which will happily eat it if they can reach it. A moisture barrier should never be used in a goat shelter because it will result in condensation and unwanted moisture inside the shelter.

Ventilation

Poor ventilation in a goat shelter can be detrimental to the goats' health and milk production. Excess moisture and harmful gases and dust can cause respiratory problems, while temperature extremes reduce productivity and make goats and humans uncomfortable. A good ventilation system distributes fresh air without drafts to all parts of a shelter, helps to maintain desired temperatures, and reduces ammonia levels.

Natural (passive) ventilation systems move air around with fixed openings such as windows, doors, and adjustable vents. Mechanical ventilation systems incorporate fans, controls, and air conditioners.

TIP: Fans and air conditioners can unnerve goats.

The noise of fans and loud air conditioners can scare or unsettle your goats so much that they do not want to be inside their shelter. It may take some time for them to become accustomed to the unfamiliar sounds. Turn on fans before you bring goats in for milking to avoid startling them.

Construction materials

Goats are just as happy in a simple shelter as they are in a state-of-the-art goat palace. A shelter can be made of stone, cement, wood, or metal. Metal is the least desirable because it is less stable, and it will be cold in the winter and hot in the summer. Do not use soft materials such as plasterboard or plywood where goats are kept because they will inevitably destroy it. All materials need to be sturdy and resistant to gnawing.

There should be no gaps, holes, or openings anywhere wider than 2 inches as noses, hooves, and even small heads could get caught when curious goats investigate. Check for open spaces between the wall and floor. Interior partitions should have openings no wider than 2 inches between slats. Make certain the floor is free of holes. The roof of a goat shelter should be designed to keep goats from climbing on top of it.

TIP: Wooden shelters

There are a few points to keep in mind when selecting wood for your goat shelter. Some woods splinter more easily than others. Goats also gnaw at certain types of wood, particularly if they are bored or if they are lacking in minerals. Be very careful about using treated wood and selecting paints or stains — some of them contain ingredients that can poison or sicken your goats. When building partitions of wood, do not leave more than 2 inches of space between slats because a horn or leg, or even a kid's nose, may get caught. Use screws for construction instead of nails, which will make it much easier to disassemble the structure if you decide to remodel later on.

Lighting

Windows and natural light inside the goat shelter encourage milk production and help with ventilation during warm weather. You should be able to close windows in bad weather. Electric lighting inside the shelter is helpful when you are milking or doing chores on winter evenings, or if there is an emergency during the night.

During some times of the year, lighting in the goat shelter can be used to increase milk production and stimulate earlier breeding. In northern geographical latitudes, increase light inside the barn beginning in September, when milk production might normally wane. Setting lights on an automatic timer to shut off for only four hours during the night, when the goats would normally be in the dark for ten hours or more, will help to keep milk production at normal levels as shorter days and colder weather set in.

Goats normally breed in the fall, when days begin to get shorter, so they are pregnant through the winter months (about 150 days on average) and give birth in spring. If all your goats are kidding at the same time, they will lactate at the same time, and be drying off (ceasing lactation) within weeks of one another. This leaves you without consistent milk production throughout the year. You may wish to influence the breeding cycles of some of your does so they give birth in fall; this way lactation cycles, lactation curves, and drying off periods are staggered and your milk production is steady. Decreasing available barn light in early spring will simulate the naturally occurring shortening of days in fall, encourage estrus in does, and breeding behavior in bucks.

Flooring

Most goat keepers prefer concrete flooring because it is easy to keep clean. It can easily be mopped with bleach. Concrete does not absorb urine, so moisture will accumulate on its surface. If you design the floor so there is a central gutter (outside the stalls of course), or if the floor slopes slightly so that urine and wash water will run toward a gutter in the side of the stall or pen, it will not be difficult to keep your stalls dry and urine free. A concrete floor is cold and hard, so you will have to be diligent about providing adequate bedding material for insulation.

Wood or hard dirt floors are softer and stay warmer than concrete. However, wood will tend to absorb moisture (and urine) and will be susceptible to rot and mold. It will be more difficult to keep clean. Dirt floors have a similar problem: urine will soak into the floor and make mud, on which your goats will lie. It will also be difficult to disinfect the floor with a bleach wash or other detergent because it will just soak into the dirt. You will not be able to disinfect a dirt floor because it is impossible to safely wash away detergents. Keep these factors in mind when deciding on flooring for your goat shelter.

Sleeping platforms

Goats like sleeping platforms — simple wooden platforms raised a few inches off the floor in their resting area. These keep goats up off a cold, hard floor and away from rodents or insects that might be active in their bedding at night. They are also cooler in summer because air circulates underneath. Make these platforms movable or with removable slats so you can clean underneath. They should be lightly covered with bedding.

Bedding

Regardless of what sort of floor you have, you must provide bedding in the stalls and rest areas so goats can find warmth and so they do not develop pressure sores from lying on hard surfaces. You cannot have too much bedding for goats. Store bedding material in a dry area before use so it is not damp or prone to mold. Straw makes the best litter, but you can also use sawdust or peat. About 1 pound of straw per goat should be added to the bedding per day.

Urine and fecal matter will accumulate at the bottom of your goats' bedding, creating a damp area underneath. As long as you have enough thickness on top of this to keep your goats dry, the damp area is acceptable and in fact unavoidable. Fresh dry hay can be spread on top of the bedding every day, a method called deep littering. In some conditions, the layers underneath will begin to compost, creating extra warmth for your goats. The deep mattress of straw bedding needs to be cleaned out about once every three months and replaced with fresh bedding to control odor and insects. Do not forget that as the straw bedding grows deeper, it raises the height of the floor, and the goats will find it easier to get over partitions and enclosure walls.

Do not let the used bedding accumulate: remove it immediately to the fields or garden, compost it, or sell it as fertilizer. Remember to use a wheelbarrow designated only for manure and bedding, not one used for hay and feed. If you are unable to dispose of used bedding immediately, you will need a manure storage area of approximately 11 square feet per goat and perhaps 0.6 cubic yards for a dung-water trench.

Feeding Area

Goats are picky eaters. They yank out selected morsels and step backward, scattering hay on the floor of the feeding area. They will not eat soiled feed or hay off the floor. Goats should not eat off the ground because they can ingest parasites from fecal matter.

Ordinary livestock feeding racks and troughs are wasteful because so much of the hay falls on the ground, and because goats often soil their food by putting their front feet in the racks or jumping up into them. Goat farmers have developed many devices for feeding goats. The best feed stations (also referred to as feeders or mangers) are designed with the nature of the goat in mind and are safe as well as practical. Not only are goats messy when eating, but they can also be pushy and bossy at mealtime; they do not like to share. An ideal feeding station is designed to minimize pushing, minimize waste, and keep food off the ground. It should have enough openings to allow all your goats to feed comfortably, and the openings should be narrow enough so that curious kids cannot jump in and soil the hay with their dirty little hooves.

Wire feeders

Many goat owners buy commercial heavy wire feeders designed to hold hay. The best of these are sturdy and take a lot of abuse without sustaining serious damage. Poorly made wire feeders are flimsy and tear easily. The resulting loose wire ends can create a puncture wound or poke a goat in the eye. Some owners will not use wire feeders because goats can easily hook a hoof in them or horned goats can become tangled and stuck. Because they are wire cages, they are not appropriate for holding anything but hay. These feeders come in a variety of shapes and sizes.

A common type fits firmly over the side of a stall; it is relatively inexpensive and several of these can be purchased to place in various locations. A wire feeder can be easily moved from place to place as long as there is a sturdy bar or low wall to hang it on. This is useful when you need to separate mothers and kids from the rest of the herd or isolate a sick goat. A variation of this type of feeder can be placed on a common low wall or bar between pens, where hungry goats can access it from either side.

You will frequently see a fence-line feeder on a goat farm. This is a feeder made of heavy wire mesh that hangs on the fence or is made up of a loose section of fence. These are frequently home-made rather than purchased commercially.

Bag feeders

At livestock shows, you will notice that some people carry portable hay bags, which can be filled with hay and hung on a bar as a temporary feeder. These bags have large openings in the sides through which the goats can access the hay and pull it out to eat. Some are made entirely of netting. Although these are commonly used for horses, many goat owners point out that these types of bags can be safety hazards for goats that tend to climb up on feeders. A goat would eventually get tangled in the bag feeder and break a leg — or a neck — trying to free itself.

Wooden mangers

A popular type of manger is the keyhole manger, a wooden structure that allows each goat a slot through which to place its head while eating. This design is useful because it discourages competition and makes it more difficult for goats to toss the hay out of the manger. Some of these feature a latch that comes down to restrain the goats in the slots.

Another type of feeder holds the hay in a rack slightly above the goats with slats sloping outward at an angle of 63 degrees. The space between the slats is narrow enough (4 inches for kids, 5.5 inches for does, and 7 inches for bucks) that the goats have to tilt their heads sideways to get them into the rack to eat. If the goat steps backwards, the slats trap its ears. Goats soon learn to keep their heads inside the rack while they eat.

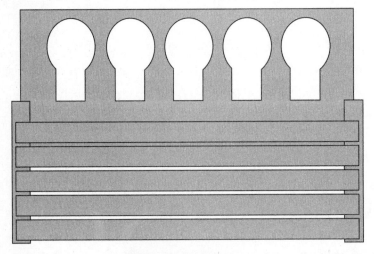

Examples of keyhole feeders

Feeding shelves

The best solution is a feeding shelf protected by a feeding gate through which the goats stick their heads to eat. A feeding station that keeps goats separated while they eat prevents the bigger animals and dominant does from depriving the smaller goats of the grain and high protein feed they need to produce milk.

The feeding shelf should be at least 19.5 inches wide and long enough to accommodate all the goats. For large, fully-grown goats, the shelf should be mounted at a height of 15.5 to 19.5 inches above the floor. Kids and smaller goats can be accommodated by building the shelf so its height can be adjusted as they grow or by placing a narrow step or shelf in front of the feeding gate about 10 inches above the floor, on which they can place their front feet while feeding. Nail a raised edge or board along the front of the feeding shelf to keep the food from sliding off.

A feeding gate in front of the feeding shelf acts as a barrier that prevents the goats from jumping onto the shelf. The goats thrust their heads through openings in the feeding gate and sort through the food on the shelf, picking out what they want to eat. The feeding gate forces the goats to spread out while feeding and ensures each goat has equal access to the feeding shelf. There are many types of feeding gates, but the most efficient ones have openings shaped like keyholes, with a round hole 7 inches in diameter at the top and a 4-inch-wide slot coming down from it. Each goat fits its head in through the round opening at the top of the keyhole and slides its neck down into the slot to reach the food. The narrower slot opening prevents the goat from drawing directly backward and pulling hay onto the floor. It also holds the goat in place while it is eating, so you can inspect it and perform simple treatments. If you want to restrain the goat, you can hold its head in the slot by securing a bar or latch across the slot above the goat's neck while it is feeding.

In addition to the feeding shelf, you can place a hayrack on the wall in the resting area and fill it with inferior hay. Anytime they want, the goats can eat hay from the rack, and whatever falls on the floor becomes part of their bedding. Also, you could suspend a hook from the ceiling to hold bunches of fresh forage. Mount mineral licks on a wall or a stand up off the floor to keep them from getting stepped on and soiled.

Many feeding stations are designed so both sides are accessible to feeding animals. In this case, there is a short wall separating the shelves or troughs, and they are designed with safety features so the bigger goats do not try to clamber over the top and injure themselves or other goats. Some feeding stations are located on the outside of a goat shelter, protected from the weather.

Plans for building goat feeding stations are widely available online or from county agricultural extension offices. Many goat farmers build their own, and some are willing to build them for other goat owners. Dairy suppliers and goat supply catalogues sell ready-made feeding stations. Regardless of which types of feeders you choose, always have safety in mind. Try to imagine all the ways that a goat can be injured before you invest in any feeder.

Watering

Goats always need a supply of fresh, clean drinking water, especially when they are lactating. For a small herd of a few goats, you can supply water in buckets or tubs. Several small containers are preferable to a single large one because they are easier to clean, and if one is soiled or spilled, there is still clean water in another container. Because goats tend to knock these over or step in water containers and soil their water, it should be changed twice a day. For larger herds, automatic drinking troughs designed especially for goats are available. In winter, special heaters can be placed in water containers to keep the drinking water from freezing.

Buck Stables

If you are keeping bucks for mating, they will have to be separated from the does. Several bucks can be kept together in an open area. Their enclosure must be fortified to withstand their strength, and partition walls should be at least 5 feet high to prevent goats from escaping. Bucks need about 33 square feet of resting space and 54 square feet of yard, preferably outside. Bucks should be kept at least 50 feet downwind from the does to prevent their odors from affecting the does' milk.

Accommodations for Kids and Mothers

If you plan to keep kids together with their mothers for several days after birth or to raise some of your kids to add to your herd, arrange an area in your shelter where the mothers and newborn kids can be separated from the other goats. Any spaces in the stall partitions must be small enough that tiny kids cannot slip through under the partitions or between slats, so make sure you address these spaces. Arrange provisions for feeding and watering the doe and her kid(s) in the stall. To give young kids access to water and feed, watering pans and feeding shelves will have to be placed lower or made so their height can be adjusted.

Kids of the same age can be successfully raised together in a group, but do not mix kids of different ages because older kids can transmit disease and bacteria to the younger ones. To avoid the spread of disease and to ensure they are getting the proper amount of nutrition, kids should be fed separately from adult goats in their own area through a feeding gate that prevents them from spilling and soiling their food. Allow 5.5 square feet of floor space and about 8 inches along a feeding shelf for each kid. After the kids stop nursing, you can move them to a young goat area where they have more space (10.75 square feet). The amount of space you need to allot for kids and young goats depends on how you plan to operate your dairy business and how large your herd will be.

When space is limited

On traditional small farms with little space for stabling, goats are kept in a "tethered stable" where each goat is tied to its manger. A space of 24 by 47 square inches is allowed for each goat. Because the goat rests and feeds in the same area, the stall must be cleaned daily and refurnished with fresh straw. To make cleaning easier, the floor should be sloped slightly toward a collecting ditch in the rear.

Pastures and Outdoor Enclosures

Most goats are kept outside during daylight hours for at least part of the year, where they can freely forage for vegetation and obtain adequate exercise, sunshine, and fresh air. Goats can be pastured together with horses, cattle, or sheep because they browse vegetation other livestock will not touch. Sheep, for example, more commonly graze on grasses and other low plants, while goats tend to prefer vegetation that is up off the ground, such as shrubs, scrub, underbrush, and even twigs, buds, and bark from trees. Owners of meat and fiber goats often turn them loose into wooded areas to browse. This is not advised for dairy goats because they can injure their udders on rough undergrowth and because the quality and type of forage they eat affects the flavor of their milk.

Dairy goats are typically confined in open pasture and kept closer to the stable or barn where they are milked. Goats like to hang around close to the shed or barn; they are less apt than sheep or cattle to wander for long distances. They also like to be close to where people are.

You have probably heard farmers jokingly remark that a few sheep or goats will clear your backyard of weeds faster and better than any professional weed control service. It is possible for three or four goats to clear a few acres of weeds in a few days. This needs to be taken into consideration as you manage the land you allot to your goats.

TIP: What goats will do to your backyard

A schoolteacher and her family have a lovely home on a few acres in a Minnesota suburb. Here they have a small shed and a large pen, and for years they have kept around half a dozen meat goats at any given time, a hobby the entire family enjoys together. Inside the pen, they have thoughtfully placed an old car, a few large wooden doghouses, and other objects for goats to climb on. The enclosure once abounded with several large, old, healthy trees with lush greenery that overhung part of the backyard. Now, these trees are stripped bare of leaves for several feet up, and they do not look healthy; craggy masses of twigs hang off trunks of damaged bark. There is really no vegetative ground cover to speak of. The goats cleared it out long ago, and what remains is a large enclosure of hard-packed dirt. The goats love it, of course, and they are thriving.

If the goats are kept in the same area all the time, your goat yard will soon become bare dirt, which turns into mud in inclement weather; remember that udders have to be cleaned twice a day for milking. The amount of forage your goats can access will affect your budget — if they are grazing on pasture, they are not eating as much feed out of your barn. In winter months, you will be providing almost all of their food, but you can lighten the financial burden in warmer months by providing access to areas for foraging and browsing.

The ideal solution is to establish at least one extra enclosure that can be left empty for weeks at a time to recover while the goats are in another pen. The size of these enclosures is up to you. A larger area will stay cleaner and give your goats more exercise. It will be necessary to maintain some control over what grows within the enclosures, such as toxic weeds. If you enjoy walking, wander over a large pasture periodically to check the quality of forage. *There is more information on plants that are poisonous or affect milk quality in Chapter 6.* The presence of a few poisonous plants will not necessarily harm your goats; if they have enough other forage to eat, they are not likely to consume more than a tiny amount of a harmful plant.

TIP: Pasture rotation

In a pasture, goats naturally seek out their favorite plants first and ignore less desirable but perfectly nourishing forage. To get the most out of the forage in your pasture, rotate your goats through smaller subsections within the pasture. They will quickly finish off their favorite plants in a sub-enclosure and then be forced to eat the other plants until you move them to a new area. Doing this will allow plants to recover in the sections that are resting, so the goats will have even more to eat when they return there.

If you do not have enough land to fence two larger enclosures, you will need to have at least one. Additionally, you will need at least one small pen to separate livestock when weaning kids or isolating a sick or overly aggressive animal from the herd. If you decide you want to keep a buck for breeding, you must be able to accommodate him separately from your does.

To move a small herd of goats from one pen to another, or from pasture to barn, all you need is a bucket of grain. Simply walk out into the field and rattle it, and the goats will follow you anywhere. However, a larger herd of a dozen goats, even fewer if you are physically small and have a large breed, will soon have you on the ground in their goat-like enthusiasm to get at the grain. A herding dog can help you move a large herd quickly and efficiently. A well-trained herding dog is also able to separate particular animals from the others, such as when kids and their mothers are reluctant to leave each other.

Goat Manure

Whether you have a pasture, yard, or grassless pen, do not allow too much fecal matter to accumulate. Allowing it to remain on the ground — even when it dries in the sun — encourages insects. Keeping your enclosures and bedding areas free of feces helps to control internal parasites (worms) in your goats because some parasite larva can live for days and months in manure on the ground. Goat manure is different from the heavy, wet manure of cows: the small, solid piles of goat pellets are relatively inoffensive and easy to collect. Manure can be collected into a pile away from the barn and other living areas. You might also add used bedding from the barn to this pile. Adding a compost activator will help it break down naturally. Many goat owners use manure for fertilizer. You will find that you can even sell it to neighbors to spread on their gardens.

Dairy Farm Office

Depending on the size of your herd and the extent of your dairy operations, you will need a place to keep and organize medical and breeding records, files, receipts and expense reports, catalogues, and reference books. This can be a bookshelf or filing cabinet in your home, but if you are constructing a barn for a larger dairy operation, create a secure room where you can have your office and secure storage for equipment and medical supplies.

At minimum, you will need space for a desk, computer, and filing cabinet or drawers, and plenty of wall space for posting lists, notes, instructions, schedules, and calendars for when you start breeding and birthing. Post a list of emergency numbers where they are easily visible. Prepare folders or files to hold receipts and information for taxes, ordering, licensing and regulation, and first aid. If possible, set up your office space outside your living area, so you can enjoy your family life or leisure time without work looming in the corner. A workshop space, a spare room, or in a room located in the barn, safe from goats that will gladly eat your tax forms, is ideal.

Here is a list of office supplies to help in figuring costs:

- A table or desk with enough room for a computer and additional space for work
- A computer you can trust, with good bookkeeping software and reliable printer
- A comfortable chair
- A large bulletin board

- A wall calendar you can write on
- A shelf for reference books and manuals (this book can be the first!)
- Address book or Rolodex®
- File cabinet or boxes
- Hanging file folders and labels or tabs
- Paper, pencils, pens, paper clips

Storage

Caring for your goats will be much easier if tools, equipment, and supplies are stored in a location convenient to your goat shelter.

Because you will be transporting feed twice a day or more, store hay and feed as near to feeding areas as possible. The amount of storage space depends on whether you will be buying your supplies in large quantities to last for several months or in smaller quantities a month at a time. Hay must be stored where it will remain clean and dry. It can be stored in an outdoor shelter with several secure sides and a roof, as long as the hay is not rained on and the goats cannot access it. Ideally, hay should be stored in a separate area inside your stable or barn. If you are buying a year's supply, remember half of it will be gone by the time your kids are born, and you can combine the hay storage area with kidding pens. Hang a serrated knife somewhere high on the wall in your hay storage area so you can easily cut twine from bales. Have a container available to hold discarded twine so the goats do not get hold of it. Twine can be a danger to the goats, as well as to other animals in the area, if it gets into their digestive systems.

You will need an area to store containers of grain, supplements, and other feed, such as hay pellets. In many dairy operations, these are in the same area as the sink and refrigerator, for convenience. Grain and supplemental feeds should be kept dry and free from rot and mold in an airtight enclosure where it is safe from vermin and insects.

Restraint and grooming equipment

Equipment for leading and restraining your goats can be stored in your feed storage area in the place where you have wheelbarrows and tools, or any other convenient place, as long as it is away from goats. Hang lead ropes, ropes for tying, and halters on the wall for easy access. Never hang ropes where goats can chew them.

Brushes

Keep a sturdy brush or two to brush dirt and debris from your goats before you milk, or before you take them visiting or showing. Get a few different types of brushes. You will find them at local tack supply shops (look in the horse supplies). Goat hair is wiry and coarse, and it can range from very thick to thin depending upon breed and individual. A curry brush can painlessly remove dirt from thin hair and condition the coat when used frequently. You may need a heavier-bristled brush for longer-haired goats. Keep a large supply of clean old towels nearby to use when you need to clean a goat, stop bleeding from an accidental injury, and for birthing.

Clippers

For grooming, you will need a good pair of clippers. Clippers cost from $100 to a few hundred dollars and run on rechargeable batteries. Most come with several different detachable blades so you can control the closeness of the shave. Clean and oil your clippers according to the manufacturer's instructions. The better care you take of them, the better they will function, and the longer they will last. In addition to grooming, these clippers can be used for such tasks as clipping away the fur around a wound so it can be cleaned or sutured. If you have trouble finding clippers, ask your veterinarian or a local dog groomer to recommend a supplier.

Hoof trimmers

Hoof trimmers are necessary to keep your goats' feet in good shape. Hooves must be trimmed back, particularly if you do not live on rocky ground where the hooves are naturally worn down. Hoof trimmers look something like garden pruning shears and do not have to be expensive to do a good job.

Tattoo gun

You will also need a tattooing device to mark your goats for identification. Young dairy goats are tattooed inside the ear, with the exception of the nearly earless LaMancha, which is tattooed on the underside of the tail. Meat and fiber goats are often given ear-tags rather than tattoos so as not to scar the flesh or damage wool, which are the marketable commodities in those breeds.

Store such items where you can easily locate them when needed. Hoof trimmers, clippers, and tattoo guns need to be in a dry, inside place, safe from weather and rust. Store each item along with its accessories to avoid losing anything just when you need to use it.

If you wish to hike and backpack with your goats, you will also want to load a few items on their backs when you halter and lead them, so the feel of weight on the back is something acceptable and normal long before you actually take them for the first hike.

Medical Supplies and Storage

As you become more experienced with goats, you will keep medical supplies on hand that you will not find necessary in the beginning, when you are more apt to rely on your veterinarian. Over time, you will be able to assess what you can and cannot do yourself without veterinary assistance, and you will discover how to save on veterinary fees.

You will need to have some medical supplies on hand right away. Some of these items are for routine health tasks and should always be available. Others are related to pregnancy and kidding and will only be used at birthing time.

General medical supplies

- Clean towels and small sponges for applying topical medication such as ointments
- Antiseptic solution, such as tincture of iodine, widely available as Betadine®
- A pair of sturdy household scissors
- Bandage scissors, surgical scissors, and hemostats (great for removing splinters)
- A few gallon-sized jugs for mixing large quantities of solutions
- Prescribed medications you are giving regularly or daily. Some of these will need to be refrigerated.
- A small box of large syringes. These are handy for flooding cleaning solution into a wound or flushing debris from an eye.
- A few oral syringes. These are large syringes for administering liquid medicines.

- A balling gun. This is a metal contraption for giving large pills; it shoots the pill quickly down the throat before the goat can spit it out.
- A set of measuring cups and spoons. These can also be kept with your feeding supplements, such as vitamin powders.
- Mineral oil
- Bicarbonate of soda (powder form)
- Corn syrup or molasses
- Worming medications (check for refrigeration requirements). Your veterinarian can show you how to administer these the first time, but once you have learned how to give your goats the medication, you can purchase and administer many varieties yourself to save money.
- California mastitis test kits
- Vaccines. By law, your veterinarian must administer some vaccines. Others, you can give yourself. Check with your veterinarian. Allow your veterinarian to show you how the first time or two because giving a vaccine incorrectly can result in infection or otherwise injure the animal. Most vaccines will need to be refrigerated for storage and will have an expiration date after which you should dispose of them.
- Styptic powder or cornstarch to stop bleeding in small wounds or cuts
- A large animal thermometer, available at veterinary supply stores
- Triple antibiotic cream for light wounds such as cuts or bleeding scratches, preferably with an analgesic (painkiller) added

- A large quantity of a good-quality udder balm. You will be using this daily for the udders of your doe, and on your own hands.
- Disposable gloves. If you have a latex allergy, you can find gloves that do not contain latex. You should have short gloves and some long ones.
- Disinfectant solution, such as Nolvasan®
- Syringes, 3cc and 12cc: These come in small boxes containing several needles, sized 18 or 20 gauge x 1 inch.

Some supplies, including a disbudding iron or caustic ointment and castrating equipment, should only be used by your veterinarian until you learn to handle them properly. *These items are discussed further in Chapter 9.*

Emergency supplies

A first aid kit should always accessible close to where the goats are kept. Some of the items for the emergency kit are also on your general medical supplies list, but you want to keep them together with your other first aid supplies so they will be on hand in an emergency. Your goat first aid kit will need to include:

- Electrolyte replacement in either powder or liquid form
- Emergency phone numbers for more than one veterinarian, and for poison control
- A stomach tube, adult goat size (livestock or veterinary supply). This can be used with your large oral syringe.
- Scissors
- Splints
- Triple antibiotic ointment
- Analgesic spray
- Clean towels

- Vet wrap bandages
- Hydrogen peroxide
- General lubricant, available at a tack or veterinary supply

Birthing supplies

Some of your birthing supplies will be found among your general supplies. A few are exclusively for kidding and should be located with kidding supplies for ready accessibility. These include:

- A stomach tube for kids and a weak kid syringe
- General lubricant
- A small (human size) thermometer, and a large animal thermometer with a string tied to it
- Sturdy household scissors
- Surgical scissors
- Molasses to put in the mother's drinking water after she gives birth to encourage her to drink more
- Oral syringes
- Electrolyte powder or liquid
- Tincture of iodine
- A hot water bottle (do not use an electrical heating pad because it can easily burn a newborn's skin as it lies helpless)
- A heat lamp hung in the kidding stall to provide a continual source of heat and greater warmth for young kids

All medical supplies should be stored well out of reach of goats and children in a dry inside location away from direct sunlight (as from a window). Do not put off stocking and restocking your medical supplies. Emergencies are never planned.

CASE STUDY: HAYSTACK MOUNTAIN GOAT DAIRY LEAVES A HUGE LEGACY

Jim Schott already had a full professional life. He had taught at public school and college and had held various other positions in the education world. But something was missing. He needed a change of atmosphere. Supported by his wife and family, Jim explored alternative means of supporting the household; they considered a children's bookstore, a restaurant, an inn. But none of these seemed to be the right fit.

Then, one day, they visited a goat dairy and the magic happened. The family had already learned to love goat cheese, and now found that they enjoyed being around the wily animals as well. So, in 1989, they bought a farm in Niwot, Colorado, and started a new life.

"A farmstead goat dairy, where we raised goats and made cheese, suited my interest in food as well as my desire to do physical work and be with animals," Schott said.

Getting started was all about commitment and educating themselves. "We bought books, attended goat shows, and talked to many people. We found a consultant from Wisconsin who helped us have the confidence to move forward with very little experience," Schott said. They befriended a veterinarian who was considered a goat expert and depended upon her for advice as they learned.

"We started with five goats, which grew to 25 very quickly when we exercised an opportunity to buy an entire healthy herd at a bargain price. We began immediately making plans for building barns and designing a farmstead dairy where we would raise the goats and make the cheese," he said.

Their oldest daughter Gretchen left her job in Massachusetts to move home and help with the budding dairy business. She lent her marketing expertise to getting the enterprise off the ground.

In 1992, their herd of 25 Nubian goats yielded the first official batch of fresh chèvre, and the real work of marketing began with Jim selling cheeses at the Boulder County Farmers' Market and other local farmers markets. With persistence, a growing knowledge of selling, and Gretchen's help, the business began to grow.

"We had learned about goats, we had learned about cheese, now we had to learn how to sell cheese. I suppose we expected that once we made tasty cheese the public would just line up at our door. They did not. It took Gretchen's enthusiastic sales style to put Haystack Goat Cheese in stores and restaurants all over the Front Range of Colorado," Schott said.

Over the past 18 years, Haystack Mountain Goat Cheese has won awards from across the country and sells to restaurants and food outlets nationwide. In 2007, they were operating the dairy farm and cheese factory with 22 employees and 125 milking does (plus several bucks who hung around and happily helped out when called upon). They kept Nubians, Saanens, LaManchas, and French Alpines. In the summer of 2008, the company was restructured as Jim approached retirement, and the dairy farm was sold.

Today, Haystack Mountain continues to be a leader in cheese, shipping orders as fast as the company can produce them. Goat's milk is obtained from a goat farm near Canon City, Colorado, about 120 miles to the south. The actual factory operates out of a building in Longmont, Colorado, from which cheese is sold directly to the public and is shipped all over the United States to complement dinner tables at restaurants and grocery store shelves.

From a love of goats, to a business that survives his retirement, Haystack Mountain has meant years of great memories for Jim Schott.

Haystack Mountain Goat Cheese is found in hundreds of supermarkets across the country. For information, see their website at **www.haystack goatcheese.com**, or call 720-494-8714.

Chapter 4

Acquiring Your Herd

here are many goat breeds, each with unique characteristics that make it more or less suitable for a particular purpose. Goat breeds are generally divided into four groups: meat, dairy, fiber, or pets. These categories may overlap to some extent, but for dairy farming, the priority is milk production. An Angora goat, known for its luxurious wool, can be milked, but its output will not be of the same quantity and quality as the milk of a dairy breed. The best dairy breeds are Alpine, LaMancha, Saanen, Nubian, Oberhasli, Toggenburg, and Nigerian Dwarf, each of which produces an impressive amount of milk of superior quality compared to meat or fiber breeds.

A number of factors will influence your selection of goats for your herd, including the breeds available in your area, price, quantity and quality of the milk, your future plans for the herd, and personal preference. You may decide to purchase an existing herd, in which case the selection will already be made for you. If you plan to sell purebred kids or show your goats, you will want registered purebred does. You can have more than one breed of goat in your herd, though if you want to breed pure lines, limit yourself to two or three breeds. Goats of mixed breed can be excellent milk producers and give birth to strong, healthy kids. If you will be selling your milk to a commercial processor, find out how they pay for milk. Some milk processors buy milk by volume, but some pay you according to the butterfat content of the milk. In that case, you want a breed that produces milk with high butterfat content. Nigerian Dwarfs, at 60 to 75 pounds, are easy for children to handle and produce high quality milk, but one or two of them may not provide enough milk for your family. Saanens and Nubians might provide you with more milk than you need for a small family. Nigerian Dwarfs and Nubians can also be bred year-round.

Finally, there is appearance and personality. You may enjoy seeing snow-white Saanens browsing in your pasture. Some people are charmed by the friendly faces of the giant-eared Nubians. The nearly earless LaMancha strikes others as either interesting looking or freakish. If you plan to show your goats, you will want to know something about conformation standards for their breed. Some aspects of the goat's physical appearance, such as her posture and the length of her teats, have a direct bearing on the way the goat milks and how efficient she is.

Goat breed vocabulary

- **Pedigree:** A document or chart with the recorded ancestry of a goat, its parents, grandparents, and so on.

- **Registration:** Documents verifying an animal is registered in the official herd book of a recognized registry organization for that breed. When it is said that an animal has "papers," it means that the animal is registered.

- **Purebred:** Goats that, according to their ancestry, fall into a breed group defined by national and often international breed standards. Purebreds may or may not have papers documenting their ancestry. For example, a breeder who keeps a small herd of purebreds but does not want the expense and trouble of acquiring papers for each one he owns may sell you a goat that is not officially registered. You have to take the breeder's word that the goat is purebred. Breeders are generally honest and reliable, but if you want to register your animal in the future, proving pedigree may be difficult.

- **Registered purebred:** A goat whose papers are readily available; it has been registered with an official registry organization, is listed in the herd book, and there is a document to prove it.

- **Grades:** Goats of mixed breeding.

- **Recorded grades:** Goats that meet a list of requirements regarding appearance and quality of milk. Buying a recorded grade is good way to obtain a high milk-producing goat that is less expensive than a purebred.

- **Americans:** A doe that is seven-eighths pure of one breed, or a buck that is fifteen-sixteenths pure. This is determined by tracing ancestry from a point at which two different breeds were bred and the offspring were then bred back into a pure line, until offspring are nearly pure but not quite. Americans offer you the opportunity to know something from their pedigree about how the goat will grow and produce, as you would with a purebred, but they are often less expensive.

- **Native on appearance (NOA):** Appears to be of a specific breed.

- **Experimental:** A term that denotes an accidental breeding between purebred goats of different breeds.

Two terms refer specifically to milking quality rather than ancestry. A doe that falls under one of these categories might be attractive. Both of these classifications are based upon a point system and are determined by someone independent of the owner.

- **Advanced registry:** A doe documented to have given a set amount of milk over the course of a year (high-yield).

- **Star milker:** A system whereby a goat is given a star rating depending upon her consistent yield (measured day by day) and the percent of butterfat her milk contains. A star milker generally must give 10 to 11 pounds of milk in a day.

Purebred or Mixed Breed?

There are various considerations and reasons for why a goat farmer would want purebred goats, but there are equal reasons for why one would not want purebred. The choice depends upon your goals, your finances, and your immediate priorities.

Having a purebred allows you to make predictions about the goat based on years of knowledge about the breed's behavior and milk production. Each individual is unique, and being purebred is not a guarantee, but purchasing a purebred goat provides some certainty that the goat will perform as expected. For example, you will know the breed typically produces milk containing a certain percentage of butterfat. This could be important if you intend to make cheeses and want your goat's milk to have high butterfat content. You can also know something about the goat's potential to produce milk. If your goal is to become a Grade A dairy operation and sell your milk commercially, being able to predict the milk output of your goats is useful in making business calculations.

You can also make predictions regarding adult size, temperament, reproduction, health, and vigor of a purebred goat. Some breeds are more prone to specific illnesses or physical flaws. Some breeds are can be bred year-round, such as the Nigerian Dwarf, and others breed seasonally. If your goats are all of one breed, you will be able to predict the adult size of any offspring.

In many cases, you will be able to get a better price for a purebred goat if you decide to sell it or its purebred offspring. If you want to show your goats, they must be purebred and have papers to prove it. Showing your goats advertises your farm and the quality of your goats, and this can be profitable when other goat keepers seek to purchase them for their own herds. If you want to make money from stud fees by keeping a buck or two for breeding, a purebred buck will bring higher fees, and one with some success in the show ring will bring even more.

On the other hand, starting your herd with purebred goats will be more expensive than starting with goats that are not purebred. If your immediate goal is providing milk for your family and learning to keep and milk goats, a goat of mixed breed will serve your purpose well. These goats are generally referred to as grades. A grade can give great milk, be highly productive, and make a wonderful pet. Many dairy herds are made up of mixed-breed goats, and sometimes this is deliberate. As with many species, mixed breed animals often display "hybrid vigor." They are healthier and stronger on average than individuals that are purebred.

Selecting and Buying Your Goats

After you have learned something about each breed's strengths and weaknesses, milk production, milk composition, size, appearance, and temperament, you can begin looking for your own goats. You will probably recognize some breeds, such as the Saanen, the Nigerian Dwarf, and the LaMancha, on sight because of their distinct physical traits. Begin by seeking out dairy goat herds living and breeding in your immediate area. Visit several of them and get to know the ranchers and breeders, who may later become part of your support network. As you explore resources and interact with breeders and sellers, keep a notebook handy and take notes. Question goat keepers about their experiences with their goats, and ask them why they prefer one breed to another.

When you are ready to purchase a goat, there are several places to look:

- **The Internet:** You can find goats advertised for sale on the Internet on classified ad sites such as Craigslist (**www. craigslist.org**) and on state Department of Agriculture websites. Try an Internet search with keywords such as "goat for sale" or "LaMancha."

- **Magazines and journals:** Many publications list breeders and have classified ad sections. Some of these are *Goat Biz Magazine* (**www.goatmagazine.info/goatbiz/**), *Goat Rancher* (**www.goatrancher.com**), *Dairy Goat Journal* (**www.dairygoatjournal.com**), and *The Goat Magazine*™ (**www.goatmagazine.com**).

- **Shows and events:** Livestock shows and events take place year-round. You will find goat owners and breeders — even goats for sale — at any of these. Visit the online sites or call various goat breed associations to find out when and where shows and events are taking place. These events, even when you are not quite ready to buy, are a good place to learn about breeds, see them up close, and question the experts. Visit the goat barn at a county or state fair. You can see some of the best breed specimens there, learn a bit about showing, and chat with owners.

- **Breeders:** Find breeders by contacting dairy goat associations and by doing an Internet search.

- **Auctions:** Be wary of buying a goat at a livestock auction. The animals being auctioned off are probably being sold for a reason. A goat may be an inferior milker, have a difficult temperament, or carry some undesirable genetic trait. Worse, the animal may come from a herd infected with foot rot or some other contagious disease that can spread to your healthy goats. Even if the goat is healthy, simply being in the auction pen is likely to expose it to disease. Occasionally, a good bargain can be found at an auction, but reserve this method of goat acquisition until you know goats better.

- **County extension office:** Your local agricultural extension office can refer you to farmers who raise and sell dairy goats in your immediate area. The staff can often give valuable advice, including current pricing in your area and what to expect once you start negotiating with a seller.

Pricing — not black and white

The price of a dairy goat can range from around $50 to hundreds of dollars. Many variables determine the price of a goat: the popularity of the specific breed; whether the goat is purebred, grade, or not registered at all; the goat's sex, age, and milk production; the individual's health and breeding history; and its conformation, or how closely an individual goat matches the ideal standard for the breed. Price can vary depending on the time of year, whether a goat has won a ribbon in a show, and whether a doe is pregnant. Finally, price varies according to geographical region.

You will not know exactly how much you can expect to pay for dairy goats until you start shopping around, but you can learn about the factors that go into pricing a goat and be able to judge whether you are being offered a fair price. If you are genuinely interested in a particular goat, do not hesitate to negotiate and see if you can work out a price that is satisfactory to both you and the seller. Many who are experienced in buying and selling goats will tell you, you can always find a purebred goat at a bargain price and a mixed breed offered for an unfairly high price. The more you educate yourself about the realities of pricing in your region, the less likely you are to lose money in a bad deal or unwise investment.

Evaluating a dairy goat

Browsing through the breed prices in your region will give you a general idea of what you will pay for a goat of a particular breed and what you will pay for a mixed breed.

When you evaluate a dairy goat, look for these three key physical characteristics:

- **Milking:** Is the shape and condition of the udder good, and are suspensory ligaments intact?

- **Body shape and structure:** Is the animal sturdy, with the bones supporting the musculature of the body, and is the body of the correct shape and size?

- **Overall health:** Does the animal show any sign of disease?

When you are examining a specific dairy goat, you will first want to know what sort of milker she is. A goat may be sweet, affectionate, and so attractive that passing cars slow to look at her, but if she does not produce milk, she does not suit your purpose. Ask permission to milk each goat you consider. If you do not yet know how to milk, ask to observe the milking. Observe what sort of temperament the goat displays during milking. Note the type of restraint used during milking because this is what the goat is accustomed to. Does the goat go into the restraint easily and stand easily for milking, or is there a fight? Is the platform raised? Does the person milking have to go into contortions to get under the goat? Is milking done by hand or is a machine used? A goat can adapt to a different restraint device or a different method of milking, but may give less milk during the adjustment period. A reputable goat seller will not object to your touching the goats and observing or performing milking.

A second major consideration is the goat's conformation — how closely it matches the ideal physical specifications for its breed. Conformation is important whether the goat is of a given breed or a mix. An animal's quality is judged by specific standards of appearance because these physical characteristics affect the animal's overall health and potential to produce milk. For example, the internal milking apparatus may not be formed correctly in a goat with double teats. This will make milking difficult, and output may be far less than that of a better example of the breed. Physical flaws can shorten the milking life of a doe and reduce its lifetime milk production, even if it is a good milker now. If the udder hangs too low, the goat may have weak suspensory ligaments and may eventually drag the udder and step on it, damaging it irreparably. A goat with a swayed back, legs that are too fine-boned, or bowed limbs may be crippled in a few years. If you plan to sell purebred registered kids for profit, conformation defects in the adult may pass to its kids and lower their market value, along with your reputation as a breeder. No individual goat has perfect conformation, but good conformation contributes to the overall health of your herd, lower veterinary expenses, and the future milking productivity of your dairy.

When viewed from the side, a goat's udder should be pendulous but not too low. Notice if the back is straight — this is called the topline. With a Nubian, the topline will run upward toward the rump, and the back will dip a bit; this is normal and acceptable conformation in this particular breed, but the backs of the Swiss breeds should be straight. When viewed from above, the goat should have a symmetrical wedge shape, narrow at the head and widening toward the hips.

1. Overall body type and condition

Start by getting a general idea of the goat's health and physical condition. Ask to see the goat's papers. If the goat is purebred, ask to see the registration and any show cards. A show card will contain notes from the judges on the goat's conformation. If the animal is not a purebred, it may be registered with another type of registry. Ask to see those records.

The goat's owner should keep barn records that will tell you what kind of milk producer the goat has been. If the herd is registered, there may also be records kept by an independent party of the production for the herd and individual goats. Remember that the yearly output for a goat is far more significant than its daily output because output fluctuates throughout the year. Finally, ask to see health maintenance records, which will supply information regarding any illnesses in the individual or the herd, evidence of testing for common caprine (goat) diseases, and vaccination records. If the seller does not have these records and expects you to take his or her word that health has been maintained and vaccines are up to date, be wary of buying. Do not let good rapport with a seller lead you into trusting too easily and buying carelessly.

2. Skin and fur

Take a close look at the goat's fur and the skin. Check for dullness, excessive flakiness, or any abscesses in the skin. Skin should be thin, soft, and a little loose over the middle section of the goat. Look for evidence of lice or mites, as this indicates the goat has not been well cared for. Skin problems can be an indication of internal health issues. Internal parasites affect the appearance of the skin.

Check that the fur is thick and consistent. Except for the Nubian, most goats do not have excessively shiny fur. It is usually coarse to the touch. As you become more familiar with goats, you will know how a goat's fur should look and will be able to spot abnormalities more easily. Make certain there are no bare patches and that the goat's skin generally looks clean and healthy.

3. Head, nose, and mouth

A healthy goat's eyes are bright and clear (Note: Goats' eyes have a rectangular pupil). Dull eyes are a sign of illness in an animal, and there should never be any mucous discharge from the eyes. The goat's muzzle should be broad with large, well-distended nostrils, indicating adequate nasal passages for breathing. The lips should be muscular because goats use them to browse for food. The jaws should be strong, and the wear of the teeth should be in keeping with the age the seller has reported.

The windpipe should be large and well developed. You can feel it through the neck, and you should observe no breathing difficulty. If you plan to show a purebred, look for a broad forehead and a jaw that is neither overshot nor undershot. Ears should be of correct shape and size. The neck should be slight and feminine in the doe and heavier and more masculine in a buck, with the length proportionate to the size of the animal. It should blend into the shoulders, widening toward the base.

4. Forelegs and chest

The forelegs should be set squarely under the goat to support the chest, not too narrowly or too far apart, as both positions will result in eventual undue strain on the shoulders and knees. Look for a broad chest, which will probably contain a well-developed respiratory system.

5. Barrel

The barrel, or body of the goat, should be broad, and deep — large in every way. This is where the digestive system resides, and a well-developed digestive system requires plenty of space to function properly.

6. Hipbones and Rump

The shape of the hipbones and rump is important because this part of the goat's body carries the weight of the udder. The distance between the hipbones and pin bones should not be too narrow but should appear to support the animal sturdily. The rump should be broad, and the slope slight. A broad rump supports an udder well, and a broad shape also indicates the suspensory ligaments are well attached.

7. Udder and teats

The udder should be firm, with strong suspension. Large size is not important: udders that are too large or too pendulous can eventually hang low and risk being injured or stepped on. A large udder has less to do with milking ability than with fleshiness. You should not see any growths, rashes, or abnormalities on the skin of the udder.

The teats should be well formed — long enough to milk and not too wide. Sometimes called sausage teats, wide teats can interfere with the function of the milking apparatus. Keep in mind that short teats often lengthen over time with milking. If a goat has short teats, allowing her kids to nurse may stretch them and make them longer. There should not be double teats or teats with two orifices (openings).

8. Wattles and horns

Some goats have them; some do not. Wattles are folds of skin under the face on the side of the neck. Usually these are cut away with a clean pair of scissors shortly after birth. They are purely ornamental and in fact can be a problem because kids and other goats sometimes bite at or suck at them and cause sores. Horns are also ornamental. Most dairy breeders remove horns because they can be used as weapons, and a goat can accidentally hurt another goat or a human with them. Some breeders prefer the decorative nature of the horns or want the goats to be as natural as possible, so they leave them intact and learn to be cautious around them.

If you plan to show your goat, horns are not permitted and should have been removed. Horns that are not removed when the kid is young (disbudding) are difficult to remove later (dehorning), and this must be done with a surgical operation performed by a veterinarian.

9. A word about weight

Excess weight puts stress on joints and ligaments and causes serious complications during pregnancy. If you wish your goat to be healthy and stay that way, it should not be overweight. A veterinarian can teach you to judge when your goats are gaining weight, in which case, cut down on feed. If you purchase a goat that is obviously overweight, ask about any current and former health issues, and get the weight off as soon as you can.

10. Look at the seller's environment

Look around the goats' home for signs that the seller is responsible and cares about the goats. Goats that have been valued and well maintained are more likely to be healthy. Check for signs of cleanliness in the barn. Is it well swept and is there good air circulation? Are the goats tied? Do they have plenty of room for exercise and good shelter from the elements? Are feeders and stalls clean? Is the water clean and fresh? If the seller's barn records are incomplete and poorly kept, chances are his or her goats have been equally neglected.

Buying Your Goats

When you purchase a goat, it is important to have a written contract between you and the seller that specifies who is responsible for what, how payment will be made, who will transport the goat, what kind of health testing will be done, and what will happen if the goat dies after the deposit is paid. Many breeders use standard contracts; if your seller does not have a contract, prepare one yourself. *See Appendix B for a sample goat sale contract.* Read the terms of the contract carefully before you sign it.

Here are some general guidelines for the purchase of your new goat:

- **Communicate clearly.** Tell the seller up front what you are looking for, and do not waste the seller's time. If you think a seller might be waiting to hear from you, and you are not interested, say so as soon as possible, so the seller can show the goats to another buyer.

- **Settle any outstanding health concerns.** If there has been a problem with disease in the herd, or you suspect a problem, request a blood test. You may be required to pay for this, but it may save you hundreds of dollars in veterinary expenses later on. A disease-carrying goat can spread a serious infection to your own herd or infect future offspring. Be firm about health concerns — a responsible seller will not mind your diligence.

- **When negotiating a price, be polite.** You may want to buy from this seller again, so do not burn your bridges by being pushy or rude. Protect your own interests, but be nice about it. Remember that "thank you" goes a long way. If you decide against the sale and walk away, doing so politely can preserve the relationship, and because gossip travels, it can protect you in encounters with other sellers.

- **Expect to pay a deposit.** This will not be returned if you back out of the sale, because the seller has spent time on you and has perhaps kept other interested buyers away for you. If the animal dies unexpectedly or if kids are miscarried, the deposit should be refunded.

- **Keep your word, and be on time.** Do what you say you will. Do not be late on the day you pick up the goat. Do not be late with payment. Old-fashioned manners are important when dealing with ranchers and farmers.

- **Arrange transportation for the goats.** It should be clear ahead of time whether the goats are to be picked up by you or are to be delivered. If you arrange to pick them up, arrive at the promised time. Time is money on a farm; do not keep the farmer from his or her chores. Load your goats, take care of outstanding business, and get out of the way. Once the goats are home and settled, contact the seller to say a final thank you and to notify him or her that the goats arrived safely. This will also give you the opportunity to ask any final questions you may have.

Bringing Your Goats Home

You have readied your property and bought your supplies. Your storage areas are stocked and organized. Now all you need are the goats, and the day has arrived to deliver them to their new home. Take a moment to consider last-minute details, so their arrival runs smoothly and is as stress-free as possible for the goats and for you. Good preparation will give you the opportunity to savor the joy of the day, instead of coping with chaos and frayed nerves. Having all your resources readily available and knowing you have already anticipated possible problems will give you confidence.

Do a last minute walk-through of your barn and enclosures. Check for fencing that may have come loose or doors not latching properly. Check that heating or cooling systems in the barn are working properly and that electricity and water are running as they should. You do not want to discover a problem just when it is time for the first milking. Set out as many of your milking supplies as you can now to prepare for the evening milking. Put hay and water out to greet the goats when they arrive in their new home.

During the last few days, you have obtained necessary medications from the veterinarian, and they are already in storage and ready to administer. You have supplements and vitamins ready, and you have posted a chart on the wall telling you supplements and amounts to be given for select individuals. You have mineral licks mounted in the barn, pen, and pasture.

Whether you are picking up your goats yourself or having someone else deliver them to you, make a final phone call to the seller the morning of the arrival to confirm that everyone agrees on the schedule, your goats are ready, and there will not be any unexpected delays. Ask the seller to run through once more with you the medications, special feed, or vitamin supplements that any individual goats are currently receiving. Write all these details down and post them on the wall where the feed is prepared. You will be tired at the end of the day and will have enough on your mind during the first milking without trying to remember medications off the top of your head.

Carry a small notepad and pencil in your pocket, along with a list of phone numbers you may need in the course of the day. These should include a veterinarian's phone number, preferably the number of the veterinarian who initially examined your goats. Your goats may become stressed during transport, particularly if the day is warm. Injuries can happen during loading and unloading or as they try to adjust to the new environment. Ensure that the goats have adequate access to water during transport. Your seller may wisely have withheld food to avoid having them become sick in the vehicle.

If you are picking up the goats yourself, do not do it alone. Take an assistant with you in case you have to stop along the way. If you are working the goats alone the first evening, recruit a friend or neighbor who is willing to come and help if you are overwhelmed. Milking will be slow the first few days until you become accustomed to the routine and gain confidence. By the time the first evening's milking is done, the animals are fed, and the equipment cleaned, you will be exhausted. Having a friend there for assistance and encouragement is a good idea.

The goal on the first day is to get the goats moved, settled, fed, and milked smoothly, so they experience the least stress possible.

Although you may begin charting your does' milk outputs the first evening, remember the amounts they give for the first few days will be off due to the stress and adjustment of the move. Within a week or two, milk production levels will be more indicative of realistic patterns in your individual does.

Transporting Goats

Most goats will allow themselves to be led with a rope attached to a collar. Kids can be transported in a large pet carrier or dog crate or on held on a towel in someone's lap. A goat can be transported in any van or pickup truck, and young goats can even travel in the back seat of a car if it is covered with a plastic sheet and old towels. Secure the goat with a rope so it does not jump around and end up in the front seat. While standing, goats can keep their balance as the vehicle sways and turns. Goats transported in a livestock trailer or in the back of a pickup truck should be sheltered from wind and drafts and given adequate bedding or padding to cushion them. On long trips, supply the goats with hay and water and stop every three or four hours to let them eat, drink, and rest.

Begin Keeping Records Right Away

Besides beginning your barn (milk production) record, keep your other records accurate and organized from the first day. Obtain vaccination and worming records, along with any medical records from the seller, as well as registration records for your purebred goats. File these in folders as soon as you receive them; do not leave them in a pile on the corner of your desk. If you are paying for your goats on the day you receive them, remember to file those receipts. Keep receipts for any rental vehicles or transport fees.

File copies of licensing and regulatory paperwork. If you can, arrange to have the bills for electrical and water usage for the barn kept separate from those for your household. That makes it easy to calculate costs exclusive to the dairy when you are doing your taxes.

Set up a software program to maintain your books so you can keep track of expenses and see whether your business is making a profit. If you do not want to use a computer, record your monthly expenses and income on a ledger sheet and keep plenty of blank sheets available so you are always ready to start a fresh sheet at the beginning of a week or month. In the beginning, you will have a lot of expenses and little or no income from your goats, but by keeping monthly records, you will be able to see where your money is going and detect patterns of spending. The better your financial records, the more you will understand your business and be able to develop it to make a profit. Careful records will help your tax preparer calculate tax deductions and protect you if the Internal Revenue Service ever audits you.

Chapter 5

Feeding and Nutrition

Proper nutrition is crucial to successful dairy goat farming. Feeding a dairy goat is not like feeding a pet. Milk production and pregnancy put heavy demands on a goat's body that must be met with good nutrition. There are various schools of thought regarding the feeding of dairy goats, from complete control of the diet to allowing free pasture grazing. Goats naturally prefer foraging, but climate and the shortage of sufficient pasture usually make it necessary to feed the goats at least some additional hay and grain. Most small-scale goat farmers settle on a combination of the two. Because no two farms and no two pastures are exactly alike, working out the best feeding program for your goats takes trial and error, as well as practice.

The challenge for a dairy goat farmer is providing the necessary nutrition for each goat with the least expense. The availability of good natural forage changes with the seasons and the weather. If natural forage or hay do not provide adequate amounts of protein, carbohydrates, minerals, and vitamins, purchase grain rations and supplements to compensate for the lack of nutrients. Good quality hay means less expensive grain rations. Pregnant and lactating does have greater nutritional needs than dry does or bucks. It is wasteful to provide more nutrition than a goat needs, but inadequate nutrition results in lower milk production and poor health.

Cattle and sheep are coarse-feed eaters; they eat grass and weeds from the ground, ingesting as much as they can, and not being very selective. Goats are more like gazelles and antelopes; they eat less food but select the most nourishing parts of a plant, such as young shoots, tender leaves, and small blossoms from among grasses, shrubs, and trees. The wild goat ancestors of domestic goats adapted from one type of feeding to the other depending on the seasons and the availability of plant growth, giving birth only in the spring when the food supply was plentiful. Domestic goats can also adapt to either type of feeding, but when they are consuming large amounts of lush green grass, they are not necessarily acquiring all the nutrients they need for optimum health and milk production. Goats that graze freely in pastures may still require grain and nutritional supplements to keep them in good health.

Calculating Your Goats' Nutritional Needs

Goats need energy — referred to as TDN (total digestible nutrients) or ENE (estimated net energy) — protein, minerals, vitamins, and water. The carbohydrates and fat in grain ration (corn, oats, and sunflower seeds) supply energy. Energy needs vary with a goat's size, age, and maturity; whether it is pregnant or lactating; weather conditions; the amount of stress it is subjected to; and the nutritional elements in the rest of the goat's diet. Milkers need a sufficient quantity of energy to continue producing large quantities of milk and maintain their body weight. A good milk producer should be given about 5 percent of her body weight in nutritional energy each day. A doe that receives insufficient nutritional energy may go into heat late, and her newborn kids may be weak. Overfeeding nutritional energy causes the accumulation of body fat, may inhibit fertility, and endangers the health of a doe during kidding.

A doe with a high milk yield needs extra energy in the first three months after she gives birth to a kid. If she does not receive adequate nutrition, her body tries to compensate by breaking down its own reserves, already depleted by pregnancy.

Protein, also supplied by the grain ration, is essential for growth, reproduction, resistance to disease, and lactation. A doe needs about .07 pounds of protein for each pound of her milk that has a butterfat content of 4 percent. The amino acid content of the protein is not a concern except for very productive milkers because goats synthesize the amino acids they need in their digestive systems. Giving more protein than needed is a waste of money because excess protein is burned off as energy or eliminated by the kidneys. Grain ration or hay that is too high in protein can cause health problems such as urinary calculi, acidosis, bloat, founder, milk fever, and ketosis. *See Chapter 10 for more information about health problems in dairy goats.*

Planning a Feeding Program

Calculating your individual goat's nutritional needs and formulating your own feed mixtures is a science. Langston University offers a free goat ration calculator (**www.luresext.edu/goats/research/rationbalancer.htm**) that allows you to enter information about your goats and recommends optimum ration mixtures. Unless you grow your own grain, want to feed your goats only organic feed, or have a large herd, it is easier to use commercially prepared goat feed.

When starting out, you can avoid mistakes by buying goat feed from your local feed store and closely following the directions on the package. You can also consult a local farmer who raises dairy goats successfully and follow his or her feeding program. Your program will vary according to the amount and type of land you have, the number of goats, and your geographical location, which will determine seasonal availability of plants for browsing. Your feeding program will change with the seasons as different growth appears and disappears in your pastures, new hay is harvested, and your goats are subjected to different weather conditions.

The appropriate diet for a dairy goat is not the same as the diet of a meat or fiber goat. Do not put your goats on a program designed for cattle, horses, or other livestock animals because the needs of each species are different. Dairy goats that receive adequate nutrients will be happy, healthy, and good milk producers. The substances that go into your goat's mouth affect not only its health, but also the quality of its milk.

A Second Look at the Digestive System

Many problems with feeding can be avoided by understanding what is happening inside your goat's digestive system. *The four compartments of the stomach in a ruminant, and the function of each one, have already been described in Chapter 1.* The upper third of a goat's digestive system is occupied primarily by the rumen and the reticulum. This is where the fermentation of food takes place. When a goat eats, it quickly fills up on the best available food, then begins to ruminate, or "chew its cud." It continually burps up masses of half-chewed food from the rumen, rechews it, and swallows it again. This continues until the microbes in the rumen have broken down the food enough for it to pass through the reticulum into the omasum, where excess fluid is removed. The bottom third of the stomach system is where heavier grains and yesterday's hay — now a heavy "slurry" — are being subjected to final digestion. In the abomasum, the last nutrients are converted and absorbed into the goat's system before the waste is carried out through the colon.

TIP: A goat's rumen is proportionally large

The rumen alone of a goat holds 4 to 5 gallons, while in comparison, the stomach of a horse — which is a single-stomached animal like a human — holds about 3 to 4 gallons of material. A large dairy goat is about six to eight times smaller than a horse. The rumen of the cow, also a ruminant, holds about 40 to 50 gallons, but the average cow weighs slightly less than a horse.

A goat needs a well-developed rumen to function properly. Roughage composed of live plants and hay will make up the major part of your goat's diet because it is an important element for keeping the rumen working and healthy. Too little or too much roughage can be detrimental to goat digestion: too little roughage decreases muscle tone in the rumen, which causes it to work less efficiently, and too much roughage can disturb the balance of organisms in the rumen that work to break down fiber. In a young, milk-fed kid, the rumen and reticulum take up only 30 percent of its stomach space. In a mature doe, the rumen takes up 80 percent of the stomach space and the reticulum takes up another 5 percent. The rumen of a young goat will not increase in size without proper stretching and development. Your kids must begin eating roughage even before they are weaned from their mothers' milk so their rumens develop well.

The rumen and reticulum do not produce enzymes but contain millions of tiny organisms — enzymes and "good" bacteria — that digest and break down the cellulose, proteins, starches, and fats in the food, and convert them through fermentation into nutrients that can be used by your goat's body. These microorganisms require a specific diet to be healthy themselves; if they are not healthy, they may over-proliferate or under-proliferate, both resulting in illness in your goat. It takes time for these microorganisms to adjust to changes in your goat's diet. It is dangerous to change your goat's feed abruptly or to suddenly feed it a large quantity of forage it is not accustomed to, such as cornstalks from a field. Changes should be introduced gradually to allow the microorganisms in the goat's rumen to adapt.

The rumen is full of gases. You can see the action of fermentation if you observe your goats closely. The side of the barrel undulates slightly as the rumen churns, and the goat belches. Your veterinarian will sometimes put a stethoscope up to the goat's left side, where the rumen is located, to ascertain whether the sounds of the rumen are healthy and adequate.

> ## TIP: Get to know your goats' rumens
>
> A healthy rumen gurgles every 45 to 60 seconds. From time to time, press your ear against your goat's left side and listen to the rumen. Listen to your goats' rumens when they are healthy and when they are sick.

Water

This book has already mentioned the need for goats' drinking water to be fresh and clean. Milk production relies on water consumption, and milk quality relies upon the water's cleanliness. The bacteria living in unclean water are ingested by the goats and eventually end up in their milk, where they affect its taste and quality. Bacteria in the water can make the goat sick if ingested. Make a habit of keeping drinking water fresh and clean.

Water helps goats control their body temperature, aids in waste elimination, and improves their digestion. The amount of water an individual goat needs is determined by temperature, the moisture content of their forage, dietary salt, the amount of exercise it gets, and whether it is lactating. A lactating doe needs to consume more water than the amount of milk she produces.

Design a way to fill water containers frequently and easily. Hoses or faucets should be adequate and within easy reach of containers (remember that hoses must be wound up and taken away after filling water containers, away from chewing goats). Place plenty of water sources around in several locations where the goats regularly spend time. Elevate smaller containers, such as pails and small tubs, off the ground where they will not be stepped in and turned over, or contaminated by feed, feces, or dirt from the ground. Wash water containers regularly with a solution of a disinfectant like Nolvasan®, chlorhexidine, or bleach, scrubbing with a hard-bristle brush, and always rinsing thoroughly. These disinfectants will kill bacteria clinging to the sides of the container.

Water in outdoor containers should be prevented from freezing. Do not force your goats to eat snow because their water is frozen. Goats like their water slightly tepid. In winter, special heaters can be placed in water containers to keep the water at an ideal temperature. Does that are not lactating and bucks can be watered once a day in winter if they are given enough warm water to drink their fill.

When does drink more water, their milk production increases. Adding some molasses to the water encourages does to drink more. Warming the water in winter and cooling it during hot weather also encourages drinking. Lactating goats that have free access to water whenever they want it produce 10 percent more milk than if they are only watered twice a day. You can install automatic watering devices that supply clean water through a nozzle whenever a goat tries to drink from it.

Water makes up more than 60 percent of the soft tissue of your goat's body and about 87 percent of goat's milk. All of the fluid lost through milking, urination, and expiration (breathing) needs to be replaced daily. A goat dies when it loses more than 20 percent of its water content.

Roughage

Roughage is highly fibrous plant material. It provides energy for milkers and young goats and gives added energy to female goats in late stages of pregnancy. When broken down, roughage provides a goat with important nutrients. Goats get this needed roughage when they browse the brush and bushes in a pasture, gobble up weeds, or eat green twigs and bark from the trees. They may get it from grass clippings, dry cornstalks (called corn stover) or turnips, parsnips, and carrots. You can usually obtain beet pulp or citrus pulp, also good roughage, from the feed store. Goats can get roughage from silage (corn and hay plants allowed to ferment in a silo). Many dairy goats get daily roughage in the form of quality hay.

Green pasture

An exclusively lush pasture (green forage) can be healthy for your goats, with some limitations. Green forage generally consists of grass and succulent plants containing a lot of water; this means it contains fewer minerals than dry food. In order to get enough minerals, your goat has to eat an abundance of green forage but becomes satiated long before it has consumed enough nutrients and minerals. For this reason, green forage does not provide all the requirements of a healthy goat diet, and it is not desirable to raise goats on green pasture alone. A lush pasture diet can cause bloat if your goats consume enough of it. *For more information on bloat, a serious medical condition, see Chapter 10.*

One solution is to restrict the time your goats remain in the green pasture. However, it will be difficult to assess how much they are eating or how much green forage they are getting as a percentage of their diet. Another way to give green forage is to do it by confinement feeding, sometimes called soiling. The goat keep-er cuts roughage from the pasture and brings it into a confined space to feed the goats. This way, the amount fed to the goats can be monitored.

TIP: A word of caution about grass clippings

If you are feeding your goats grass clippings from your lawn, consider the fertilizer and weed killer content in the grass. Fertilizer is never a good thing to feed to goats, and weed killers are toxic. If you do not use fertilizers and weed killers, you can feed the grass clippings to the goats.

Mixed pasture

Pasture filled with a variety of grasses, weeds, brush, and woody plants is ideal for your goats. This is the type of pasture referred to when goat owners talk about natural, or free-range, feeding. Many dairy owners maintain a few fenced pastures. Goats benefit from the exercise and the mental stimulation of exploring a pasture.

Pasture feeding can be expensive in the long term. You may have to purchase less roughage, but an acre of land requires more than 800 feet of fencing on average, and to confine goats you will need quality fencing.

Ideal pastures are soil-tested, fertilized to compensate for defi-ciencies, planted with specific desirable grasses and plants, and managed to prevent overgrazing. Weeds that are particularly suitable for goat pasture include yarrow, daisy, chicory, dande-lion, plantain, nettle, thistle, and wild roses, such as multiflora.

Goats can clear a pasture of all growth in an amazingly short time (sometimes a day or two per acre). They eat their preferred forage first before moving on to less appealing plants, and they can end up eradicating their favorite plants entirely. The solution is to limit their time in the pasture or rotate the goats through several enclosures, or paddocks, to give plants time to regrow. This is best done by using movable fences within a large pasture enclosure with a secure, permanent fence. Because fewer of their favorite plants are available inside the smaller area, the goats finish them off and move on to other, less desirable plants. Goats can also be tethered and moved from one area to another. Tethered goats should be kept under close observation because their grazing habits make it easy for them to become tangled up in the tether and even hung on tree branches.

A pasture rotation plan takes into account the plants that grow in the pasture at different seasons, as well as the effects of rainfall or drought. Milk production will decline in goats that are left too long in a sparse, dry pasture, and their health may suffer. Goats must be supplied with drinking water and shelter from inclement weather in each enclosure. An ideal layout would allow the goats to return to the barn when it rains.

Pastureland must be monitored closely and frequently for the presence of toxic weeds. Toxic weeds can make your goats ill, do permanent damage to their nervous systems, and can even be deadly. Goats tend to avoid many poisonous weeds, and they nibble here and there, never eating a whole lot of anything at one time, so a few toxic weeds in a pasture should not be a problem. It is not true, however, that goats avoid eating toxic plants. They browse the best of what is available; if the good stuff is scarce or gone, they will turn to the toxic options. They may also accidentally ingest toxic weeds while eating something else, such as new spring grass in a low-lying area.

TIP: Poisonous plants

Countless plants contain material toxic to goats if enough of it is eaten. Some of these plants are well known, some are rare, some are poisonous only under certain circumstances or only if specific parts of them are eaten, and some are common garden ornamentals. Toxic plants are categorized by the type of poison contained, the effect of their toxins, or the part of the plant containing the poison. Some plants may contain several poisonous parts. You can find numerous lists of plants toxic to goats on the Internet (see the list maintained on the Fias Co Farm website: **http://fiascofarm.com/goats/poisonousplants. htm** and the U.S. Department of Agriculture website: **www.ars. usda.gov/Main/docs.htm?docid=10086**). It is important to know the toxic plants in your region, where they are found, when they are poisonous, and what they look like. Consult your agricultural extension office and other local goat or sheep farmers to learn about the toxic plants in your particular region, as well as recommended treatments for plant poisoning.

The severity of poisoning from a toxic plant is determined by the amount eaten, moisture levels, the health of the animal, and its age and size. Some goats can ingest quantities of a toxic plant and show no symptoms while others die. The main cause of plant poisoning is starvation; when animals cannot obtain enough regular forage, they begin to eat plants they would normally avoid. Sometimes the toxic plant is eaten by mistake, such as when white cockle, a non-native weed common in fields of alfalfa, clover, and small grains, is present in large quantities in grain fed to goats. Even when they are well provided with good feed, goats may browse on toxic plants because they are bored with their regular forage or reach over a fence to eat ornamental plants in a garden. Ask your neighbors not to feed yard clippings to your goats without asking you first.

Some common pasture plants that are toxic to goats:

Amaryllis
Avocado (leaves and fruit)
Bracken fern (grows in shade)
Dock weed
Hemlock
Locoweed
Milkweed
Mountain laurel
Oak leaves (some types only)
Sorrel
Rhubarb
White clover
Wild cherry (wilted)

Be aware of the possibility of nitrate poisoning. Many plants that are good for goats provide nitrates, which is a necessary nutrient. However, under certain conditions these healthy plants can come to contain toxic levels of nitrates. Unusually high levels of nitrates accumulate in some plants, including oats, corn, alfalfa, pigweed, lamb's quarters, and Johnson and Sudan grasses, when they undergo a growth spurt after a dry spell. Another common source of nitrate poisoning is the consumption by goats of animal waste or fertilizer. This usually happens when runoff from a nearby, contaminated field waters the plants they eat, or the goats drink it.

Hay

Hay is made from a variety of plants cut when still green and then allowed to dry (cure) before they are baled for storage. When these plants are cut at the right stage of growth and cured using the right methods, they retain nutrients that are valuable for your goat in addition to providing the fibrous material to keep its digestive system working well. Hay is more or less bright green in color and is different from straw, which is the golden-colored dry leaves and stems left after the harvesting of grains such as wheat, oats, and barley. Hay is used as feed while straw is used as bedding.

The average goat eats at least 4.5 pounds of hay per 100 pounds of body weight every day. A well-built Swiss milker will eat about 7 pounds of hay every day, and more during pregnancy. Hay that is fresher and composed of high-quality grasses and plants is more expensive. Some people make the mistake of feeding poorer-quality hay, reasoning that it saves money, but feeding high-quality hay means you will be feeding less grain and other supplements; you do not save money by using cheaper hay. Like grain, high-quality hay is high in protein, so, if your goats are getting the protein from the hay, they will not need as much in the form of grain.

Various factors affect the quality and content of nutrients in hay. A hay field may be cut, or cropped, several times during a season. The first crop is coarser and less nutritious than later cuttings. The hay must be cut at just the right time. Some hay fields contain various kinds of weeds that affect the quality and price of the hay.

The method used for curing the hay is a major factor in its quality. If it is stored too long, or left in a field curing in the hot sun for too long so that it becomes bleached, a large part of its nutritional value can be lost. Sometimes the stems are crushed to allow for more uniform drying, but nutrients are lost in this way as well. If hay becomes wet, either from being baled while damp or while lying in stacks that get rained on in the field, it can become moldy and unhealthy for your goats. Although mold in hay is often not apparent to the naked eye, after you become experienced in what good hay looks and smells like, you can usually smell moldy hay. Generally, fine-stemmed hay is higher in nutrients than coarse hay.

Hay falls into two basic categories:

- **Legume hay** is made of plants like clover or alfalfa. It is higher in protein and calcium — important nutrients for all goats but especially for lactating does. Alfalfa is considered ideal hay for dairy goats because it is rich in calcium and its protein content is around 13 percent, compared to 5 percent for some grass hays.

- **Grass hay** is made of a grass such as Timothy, Johnson, brome, or orchard grass. If your goats are eating grass hay, provide supplemental protein and calcium with grain.

Do not assume legume hay is high in protein content just because it is legume hay. These plants must be cut at a specific stage of growth; if this time is missed, nutrient levels fall off rapidly, and the resulting hay is inferior in nutritional quality.

Experiment with various types of hay and with supplementing high-protein grain. Hay — or comparable roughage — should make up the major part of the goats' diet, and their diet must contain adequate protein. Lower quality hay can be fed to adult bucks or to does that are drying off between lactations.

Hay pellets can be purchased from feed stores in various sizes. These are ideal for goats because goats tend to be messy eaters. They will rummage through a big pile of hay, strewing it all over the floor, just to find a little piece of leafy stem. If you feed pellets, there is far less mess and more hay ends up in the goat. The only disadvantage is that pellets are expensive.

Prices of hay vary according to geographical region, content, and the effect of recent weather on the harvest. To determine pricing in your area in a particular month, just shop around. It is a common procedure to have a hay sample tested for quality and content. Your local county extension office will be able to tell you who in the area can do this testing for you. Forage testing helps determine what kind of grain ration is needed to supplement a particular hay crop. If the hay is of good quality, you can use a less expensive grain mixture.

Grain: a good thing in small quantities

The amount of energy provided to the goat per unit of feed is measured in therms. Alfalfa hay, the hay probably highest in nutrients, contains about 40 therms per 100 pounds of hay. The grains corn and barley contain almost twice this amount of therms. Goats need many of the nutrients present in a good grain ration that are not present in a roughage diet.

Grain ration contains added nutrients, such as salt (needed in lactating animals), protein, fats, and added vitamins and minerals like iron. It also contains molasses, which acts as a binding material, cuts down on dust, and provides some additional minerals. Grain ration has as its foundation an assortment of grains such as corn, barley, rye, wheat bran, oats, sunflower seeds, and soybeans. It might contain soybean or linseed oils as added fat sources. It might also contain dried vegetable matter such as tomatoes, kale, carrots, turnips, beets, and parsnips.

Do not give grain ration formulated for cattle or horses to goats because the animals have different needs. For example, though goats need some molasses, the amount of molasses in horse grain ration can interfere with the digestive processes of a goat. Horse grain ration also contains a level of copper that can be toxic to a goat.

Grain ration for goats can be obtained by the bag at a feed store. When you buy it premixed, you are assured that your goats are getting the necessary nutrients, premeasured in the best proportions.

You will find that your goats are fond of their grain ration. Feeding too much of it is detrimental to their health, but they might not agree with you. You can usually use grain to lure a goat into a truck, into a stanchion, or anywhere else you need it to go. Grain must be stored securely where the goats cannot access it.

TIP: Grain feeders

Grain can be fed in a manger, in a hanging bucket, or in a feeder made of PVC pipe. This material is durable and can be cleaned easily, which makes it ideal as a container for messier grain mixes.

Important vitamins

If your goats are being fed a combination of forage and grain ration, they are probably getting the vitamins they need. Sometimes lactating or pregnant animals need more, especially if they are stall-fed and do not have the opportunity to consume a wide variety of plants through foraging.

- **Vitamins C, K, and the B:** These vitamins are all manufactured within the goat's body and do not need to be supplemented. Lactating animals may need supplementary vitamins A, D, and E.

- **Vitamin A:** Vitamin A helps goats resist various diseases. It is closely associated with reproductive health; it keeps udder cells healthy and helps the udder resist infection. It also contributes to eye health. To some extent, a goat manufactures vitamin A itself when it consumes foods that contain carotene such as carrots, yellow corn, and some green forage plants. However, vitamin A deficiency is not uncommon.

  Kids deficient in vitamin A are prone to respiratory illnesses, often exhibiting watery eyes and mucous nasal discharge, coughing, and diarrhea. Because goats as a species are particularly susceptible to pneumonia, these kids can end up with pneumonia. In adult goats with vitamin A deficiency, you might observe susceptibility to infections, or night blindness (they will panic and flee if you approach them in the dark). It is believed that vitamin A deficiency is a cause of infertility in bucks.

- **Vitamin D:** Goats produce vitamin D when exposed to sunlight, as do other mammals, and so goats need plenty of sunshine to be healthy. Vitamin D is necessary for the absorption of essential minerals, such as calcium and phosphorus.

- **Vitamin E:** Vitamin E is particularly important to the quality of milk. Vitamin E deficiency in a milk goat can cause her to produce milk that tends to spoil rapidly.

Mineral supplements

Novice livestock owners are often confused about whether and how to supplement minerals. Oversupplying some minerals can cause severe health issues that are just as dangerous as mineral deficiencies. Most goat owners simply provide supplemental minerals in loose form (available at a feed store), and make them available to the goats at all times. You will find that individuals who need supplementation to their diets tend to take just what they need.

Mineral deficiencies are often due to the composition of the soil in the region. If soil does not contain enough phosphorus, for example, there will not be phosphorus in the plant life. Or there might not be enough iron in the water. It helps to be aware of any major mineral deficiency in your area, so you can troubleshoot possible health problems when they begin to appear.

Salt blocks with added minerals are generally not favored for goat owners for several reasons. Chief among these is the tendency for the goats to overdose on salt (salty milk taste) and the tendency of the goats to climb all over the blocks and contaminate them. It is better to provide loose minerals in an indoor feeder.

Goats need relatively large quantities of calcium, phosphorous, magnesium, sodium, chlorine, potassium, and sulfur and small amounts of many trace minerals, most of which occur naturally in their feed.

- **Calcium:** Calcium is needed for bone development and growth and is secreted in milk. A lactating doe needs about .022 ounces of calcium per pound of 4 percent milk. Alfalfa and calcium supplements are a good source, but grass or mixed hay and pasture are not.

- **Phosphorous:** Phosphorous is needed for tissue development and bone growth and should be given along with calcium in a ratio of 1:2 (1 part phosphorous to 2 parts calcium). Decalcification of bones and swelling of facial bones results when too much phosphorous is fed in relation to calcium, something that often happens when goat keepers increase grain rations for young goats to try to make them grow faster.

- **Magnesium:** Magnesium is essential to the proper functioning of the nervous system. In regions where the soil is naturally low in magnesium, goats should be fed alfalfa hay or a magnesium supplement.

- **Sodium and chlorine:** A lactating doe loses as much as 1 ounce of salt (sodium chloride) for every gallon of milk produced. Salt should be freely available to your goats.

- **Potassium:** Roughage-based diets usually contain sufficient amounts of potassium, but it can be lacking in diets heavy in grain or weathered hay.

- **Sulfur:** Sulfur, a necessary component of the collagen that helps to form bones, tendons, and connective tissue and a vital element of keratin (the chief component of hair and skin), is present in most goat feeds. Soils in some regions are deficient in sulfur. Check with your agricultural extension office to see if you need to give supplemental sulfur.

- **Copper:** Copper is necessary for goats but toxic to sheep. Avoid mineral supplements labeled "for goats and sheep" because these will not contain copper.

- **Selenium:** Goats need trace amounts of selenium, but high levels are toxic to goats. Hay and grain grown in some regions of the United States contain toxic levels; while in many eastern regions, the soil is deficient in selenium.

You might want to consider providing a few minerals individually. Many goat owners claim that baking soda keeps the rumen functioning well and prevents indigestion. This can be added into the loose minerals you provide or offered separately to the goats, which will take it when they need it. Probiotics made of yeast or yeast mixtures are also believed to aid digestion and improve rumen function. They can be purchased in dry or wet form. Some goat owners add them routinely to the grain ration, while others only provide it to goats suffering from digestive problems or under stress.

Some supplements can be given as a bolus (a large round mass fed orally all at once) or by injection. These are given when there is a known deficiency of a given mineral in your region of the country.

Your veterinarian or agricultural extension officer can advise you if there is any supplement you should be giving routinely.

Bucks are prone to developing urinary calculi (bladder stones), the usually unavoidable result of an accumulation of salts. This can be quite painful, and if the stones are big enough to cause urinary obstruction, the condition is life threatening. Adding a little ammonium chloride to the buck's feed will help prevent this condition by acidifying the urine and thus preventing the stones from forming so easily. Take care though not to overdose; 1 teaspoon daily per 150 pounds of weight is plenty.

TIP: Are your goats eating well?

You can judge the success of your feeding program by observing your goats:

How long and how often does a goat eat?

What feed does it like? What feed does it avoid?

Is its rumen well filled?

Does the goat chew its cud?

Has the goat gained or lost weight over the last few weeks?

Has its milk production changed?

Does the goat's coat look healthy?

Does the goat seem energetic?

Are the goat's droppings formed into hard little balls?

The very young and the very old

It makes sense to ensure that kids are getting enough food when the herd rushes to eat, jostling each other to get at the hay. Often, feeders are designed so the little ones can eat at a lower level while the taller goats lean over them to get hay higher up. Or the kids may enjoy picking up the hay that has fallen to the ground as the older goats eat. Some goats will be more assertive than others at feeding time; the more docile animals get shoved aside and may remain hungry. Observe your smaller kids carefully and note whether they are putting on weight and growing at the rate they should.

An elderly goat may also have trouble getting adequate feed. The aged goats are weaker than the younger ones and also get pushed aside or bullied away from feeders. Sometimes, stiff older joints make moving fast enough to the feeding station a difficult task.

To ensure your older goats are getting enough nutrition, you might want to feed them in a place protected from the others, where there is not such stiff competition at mealtimes. The teeth of an older goat are well worn, which makes chewing more difficult. Special feed supplementation may be necessary to ensure malnutrition does not become a problem. Your veterinarian can help you monitor your elderly goats for nutrition deficiencies.

Summary of feeding basics

If feeding your goats seems complicated at first, do not panic. You will soon learn the basics, and over time, you will come to understand the needs of individual goats just by watching them. In the beginning, follow these basic guidelines:

- Feed your goats all the hay they can eat. If it is high quality, give less grain ration.

- Feed your goats 1 pound daily of grain ration per goat. For lactating animals, feed 1 pound extra for each 3 pounds of milk produced.

- Vary the composition of their feed.

- Allow your goats enough time to feed.

- Stress in your goats results in less efficient functioning of the rumen.

- Make necessary changes in feed gradually.

- Each goat is an individual and has different nutritional needs. Watch your goats from day to day, and you will soon learn individual needs.

- Ensure that the young and the elderly are getting adequate portions of food.

- Feed mineral supplements in loose form freely.

- Plenty of fresh, clean water should be available at all times.

Chapter 6

Milk and Dairy

Many people who taste their first glass of goat's milk comment that it is just like cow's milk. Occasionally someone's first taste is tainted for one reason or another and the taster, experiencing an odd flavor, assumes it is inherent in goat's milk. It is not. By being aware of the factors that affect the taste of goat's milk, a dairy manager can take steps to ensure its good flavor.

Goat's milk that is fresh and has been carefully obtained should taste identical to cow's milk. However, the content and structure of goat milk differs from cow milk, causing it to be nutritionally different as well. You will find many charts in books and on the Internet comparing goat's milk to cow's milk. The exact components of cow's milk and goat's milk vary according to the specific herds tested, the individuals tested, and the conditions under which testing was done. However, these charts are helpful in understanding how goat's milk differs from that of other mammals and appreciating its unique nutritional value. Goat's milk is about

87 percent water and 13 percent milk solids. The protein contents of goat's milk and cow's milk are about the same. Goat's milk is higher in vitamin A and some B vitamins, while cow's milk is higher in others. Goat's milk contains higher amounts of some minerals than cow's milk and lower amounts of other minerals.

Many people who cannot digest cow's milk find that goat's milk does not bother them. Hospitals have used goat milk for years. Physicians have long known that it is a valuable alternative, not only for persons with milk allergies, but also for weaning infants with intolerance for cow milk. There is no scientific explanation for why many people with milk intolerance can easily digest goat milk. Scientists speculate that it has something to do with either lactose, a type of sugar found in all milk, or casein, a protein that, in addition to some others, causes fat globules to clump together. Cow's milk is somewhat higher in lactose, which is believed by some to be the direct cause of milk allergies or milk intolerance in some people; these people frequently find goat's milk easier to digest. Another reason why goat milk might digest more easily is that the fat globules in goat milk are smaller because it contains only trace amounts of casein compared to cow milk.

People with skin diseases such as eczema seem to benefit from drinking goat's milk, though doctors are not sure why. Because the fat molecules in goat milk are relatively smaller and easier to digest, it is often preferable for persons with chronic liver disease as well. Dairy goat farmers view hospitals as a potential market because physicians would probably prescribe goat's milk more frequently if more fresh goat milk were readily available. However, federal regulations governing the food used in hospitals are prohibitive for small dairy operations.

A thriving cottage industry has grown in recent decades around the marketing of soaps and lotions made from goat's milk. Historically goat's milk baths have been considered herapeutic for the skin. Cleopatra bathed regularly in goat's milk. Like other types of milk, goat's milk contains many minerals, acids, and enzymes that nourish dry skin. It also contains something no other milk does: capric-capryllic triglyceride, a nutrient that makes the soap soft and your skin soft as well. Goat lotions and soaps have not been shown to cure any particular skin disease, but sufferers of psoriasis, eczema, and various skin rashes, have reported relief when using it regularly.

Homogenization and health issues

If you have ever seen butter or cream made from fresh cow's milk, you understand how milk separates; the creamy substance forms a layer on top, which is scraped off and used for the butter churn. The process of homogenization occurs when the large fat molecules break down to resemble other molecules in the milk and discourage separation. If you leave a glass of homogenized whole cow's milk on the counter for several hours, a skin of cream will not form on the top. Goat's milk is sometimes referred to as "naturally homogenized" because the fat separates out less easily than in cow's milk. The fat globules of goat's milk are smaller. In addition, goat's milk lacks a protein, called euglobin, which causes the fat molecules in cow's milk to adhere to one another and allows the cream to separate and rise easily to the top.

Homogenized cow milk might be linked to heart disease

Some researchers suggest that the process of artificial homogenization of cow's milk is linked to the sudden increase of diabetes and atherosclerotic heart disease in the United States after the 1940s, when homogenization became a widespread practice. During the 1970s and 1908s, Kurt A. Oster released research suggesting that the enzyme released into milk when fat molecules are mechanically broken enters the body through intestinal walls and can damage the heart muscle with scar tissue and lead to hardening of the arteries. His theories are still being actively debated along with other hypotheses explaining the link between homogenization and disease. If they are true, goat's milk, with smaller fat globules that are not artificially broken down, is a healthier alternative. No such heart and arterial damage has been shown to be associated with non-homogenized milk, whether from a goat or cow.

The Importance of Sanitation in Dairying

One reason for keeping dairy goats is to have a supply of milk that is healthier and more wholesome than the cow's milk sold in grocery stores. Your milk will not be wholesome if you allow it to become contaminated with bacteria. The way you handle your goats, what they eat, and how clean you keep their environment and your milking area determine the safety and palatability of your milk. Even a minute amount of dust, hair, hay, or bacteria in your milk can alter its flavor. Milk inside a healthy udder is relatively free of bacteria. From the moment your animals are milked, the likelihood of bacterial contamination of the milk can be lessened through the regular practice of some simple hygienic habits. Owners of dairy livestock perform daily tasks, such as sweeping out and hosing down the milking area, to ensure the safety and quality of the milk their animals produce.

When working with your goats, consider all the ways you can prevent bacteria from proliferating and spreading. Besides being the culprit that causes milk to have a bad taste, bacteria is responsible for transmitting disease from goat to goat and from goat to human. Good sanitation and safety habits, such as hand washing, keeping equipment clean and sanitized, and religiously dipping teats every time you milk, will prevent health problems among your goats and ensure the profitability of your dairy.

Even in the cleanest of dairies, you will occasionally have a case of mastitis. Check udders for injury and disease daily. *Mastitis will be discussed further later in this chapter.* If you find any indication of mastitis, contact your vet, and begin treatment so that a small problem does not become a big one.

Pasteurizing Milk

Many dairy goat owners keep dairy goats because they want to provide raw, unpasteurized milk for their families. However, diseases can be transmitted from goats to humans through raw milk. To rule out the possibility of transmitting disease through raw milk, have a veterinarian thoroughly inspect your herd. Keep your herd isolated from other herds or other livestock that might carry disease and be cautious about quarantining new goats and goats that have been taken to shows or fairgrounds, or even to be bred on other farms. You can have your milk regularly tested for infectious agents by a veterinary testing agent. A veterinarian can tell you where to have your milk tested.

To pasteurize milk, heat it to 185 degrees Fahrenheit (85 degrees Celsius) or heat it to between 159.8 F and 165.2 F (71 C to 74 C) for 40 seconds. Infectious agents in milk are killed when hard cheeses are ripened for a long time but not in fresh cheeses.

Milk Yield

You will be milking each lactating goat twice daily. Folk wisdom has long dictated that milking be done religiously at 12-hour intervals, but recent research in France has shown that dairy goats produce just as well if there is an eight-hour and a 14-hour interval between milkings each day. Apparently, the length of the interval between milkings is not as important as keeping a consistent schedule. Having a full udder is uncomfortable for a goat; do not make your goats wait so long to be milked that it causes them to suffer. They will let you know with loud cries if they have waited too long. Always allow at least eight hours between milking sessions, and commit yourself to milking at approximately the same times every day. Irregular milking will adversely affect your milk yield and cause stress to your animals, which can affect milk quality.

It is difficult to say exactly how much milk a goat will produce because of the variance between breeds and within a breed, and between individuals. Age, climate, weather, illness, and stress can greatly influence an individual goat's milk production. It is safe to assume that a dairy goat will give, on average, about 1,500 pounds of milk per year. Note that goat's milk is measured in pounds weight, rather than in volume. Eight pounds of milk is about a gallon, but milk is measured in pounds weight rather than volume because volume measurement is far less reliable. Fresh milk tends to foam, and a gallon jar of foamy milk contains less than a gallon jar of milk that has settled. Weighing the milk gives a more accurate indication of how much milk is in a given container.

Assuming that each goat gives about 1,500 pounds — or 187 gallons — of milk per year does not mean you can divide 1,500 by

365 days to arrive at a daily average milk production for your dairy herd. An individual goat's daily output varies from day to day. Most goat breeds give birth (also called kidding) once a year. Shortly after kidding, the adult female "freshens," or begins to give milk. Then the daily output gradually increases until the individual's production peaks. This generally occurs about two months after freshening, but varies with the individual. After peaking, the daily output will taper off again. The average goat will give milk, more or less, for 305 days of the year. Over time, the practice has evolved of allowing the individual to kid, then milking her for ten months, and then letting her dry off — giving her a rest from the milking cycle — for two months, during which time she is bred and prepares to kid and begin the cycle again.

Changes in diet and climate cause seasonal changes in milk production. In order to have a consistent, year-round milk supply, it is necessary to maintain a herd and manage the does so that at least some are always lactating.

TIP: Basic milking supplies

Milking equipment can be as simple as a halter, a tether, a stainless steel bowl from your kitchen cupboard, and some plastic funnels, but because you will be milking your goats twice a day, every day, your life will be easier if you purchase specialized milking equipment. The cost of specific items needed for milking and processing varies considerably depending on whether you milk by hand or with a machine, the current demand for these supplies, and your geographical location. When you first establish a small dairy herd, you will probably milk by hand. If you decide to increase your herd, you may want to invest in mechanical milking equipment that allows you to milk many animals quickly.

Before bringing your first goats home, acquire the following supplies for milking:

- A milk stand or a stanchion, also called a "head gate." This is a restraint device that you will use for milking and occasionally for feeding, trimming hooves, or during veterinary procedures. The stand is often elevated to hold the goat at a comfortable height for milking. A milking stand can be purchased from a dairy supply outlet or a goat catalogue, or you can build one yourself. Plans for building milking stands are available online and from agricultural extension offices.

- A milking stool. You will need more than one if several people will be milking at the same time.

- Glass containers for milk storage. Large jars are made for this purpose.

- A stainless steel milk strainer, with extra disposable filters

- At least one seamless stainless steel pail for milking

- A second stainless steel pail to be used exclusively for washing udders

- Towels and sponges for cleaning udders. These must be kept separately from barn cleaning supplies.

- Udder wash, also called "dairy soap." This is a sort of disinfectant soap made specifically for safely and efficiently washing the udders of milking animals.

- Teat dip

- A strip cup to test for mastitis, which you will be using regularly from the first time you milk

- A supply of mastitis testing kits, such as the California Mastitis Test (CMT)

You might have an individual goat that gives a lot of milk after kidding and peaks with a high daily production, but then she dries off early. Another doe might not give the same impressive daily quantity of milk but produces milk consistently for at least 305 days before drying off. If milk production of these two individuals is compared, they might average out about the same, or the second individual might prove to be the better producer, even though her initial output and peak is not as high.

In order to better understand and predict the outputs of individual does, keep a chart of production. After milking, carefully weigh and record the yield of each goat by pounds weight to 1/10 pound. This practice eventually reveals the daily pattern and lactation curve of each goat. It also provides steady documentation of the quality of each animal as a milker, if you need the information later when selling the goat or applying for special recognition or status. The output of an individual doe will vary from year to year as the animal matures and ages. Over time, you will learn how to stagger the pregnancies, kidding, and drying off of your goats, so they are not all drying off at the same time. The goal is to always have a steady milk supply available.

If a goat fails to dry off naturally after ten months, and you want to give her a rest and breed her again, help her dry off by decreasing milkings to once a day for a week, then every other day for a week, then stopping at the third week. Some swelling of the udder will probably continue. This pressure actually helps stop the milk flow. If the doe does not seem to be drying off or is uncomfortable, milk her out a week after the last milking.

Abruptly stopping the milking cycle of a doe may damage the udder. Be patient and help her dry off gradually. In addition to decreasing frequency of milking, decreasing her grain ration will encourage drying off. Many goat owners also decrease water, but there is a risk of stressing the kidneys any time you deny an animal free access to plentiful fresh water. Time, patience, and giving lower rations of protein-rich feed, such as grain, will allow the doe to dry off naturally.

Udder care

Healthy udders are essential to the success of your dairy. A healthy goat has a healthy udder, and a healthy udder functions optimally to produce high-quality milk. The udder of a goat has two sides, divided by a cleft in the middle. There should be one teat on either side. The udder ideally is set high and is wide rather than being low and pendulous. It is supported by a series of internal ligaments attached to the abdominal wall. In older goats or goats that are heavy milkers, these ligaments can become stretched, allowing the udder to hang low. A low, pendulous udder is more prone to being kicked or stepped on by the goat, or injured by underbrush or other objects on the ground. In severe cases, an udder support can be put on the goat to make her more comfortable, protect the udder from injury, and prevent ligaments from further stretching.

Monitor the condition and appearance of your goats' udders daily. Check frequently for scratches, cuts, and other wounds, and watch them closely for signs of infection. A sizeable cut on the udder will often bleed profusely. Apply pressure to the cut with a clean towel until blood clots, bleeding stops, and you can better see the wound to examine it. Apply iodine immediately to discourage bacterial contamination of the wound. You may also want to apply a triple antibiotic ointment. A deep wound may require suturing, which should be done by a veterinarian or a veterinary technician to avoid infection.

Cuts or other wounds directly on the teat can be more serious. These can take time and effort to heal because every time you milk they will tend to tear open again, and this is difficult to avoid. Each time a wound opens, healing is interrupted and bacteria is again introduced into the wound. Any such wound must be kept clean and treated with care. If a teat wound leaks milk or seems to be infected, seek the help of your veterinarian.

Mastitis

When you milk, examine the udder of your goat closely for signs of abnormality. Look for wounds, redness, lumps, discharge, unusual swelling or heat, and any sign that something is not as it should be. Mastitis is a common inflammation of the udder generally caused by poor sanitation practices, insect bites becoming infected, or injury to low-hanging udders. Mastitis is the result of infection by any of several bacteria, and a simple test of the milk reveals the specific cause of mastitis in your goat so it can be treated. Prompt and aggressive veterinary treatment is important; neglected mastitis can scar the narrow passageways within the udder, do permanent damage to the teat duct, and create pockets where bacteria remain after treatment, ready to start up another infection. In the worst cases, the teat becomes gangrenous and the disease fatal.

Mastitis can be acute (sudden and severe symptoms) or subclinical (little or no obvious symptoms), and can become chronic (recurring and difficult to eradicate). Once it is determined that mastitis is present in an animal, the intervention of a veterinarian is required to determine which bacteria is responsible and the best treatment to stop it.

Some goats with mastitis will display obvious signs of illness. They may develop a fever, appear listless or lethargic, avoid being milked, or display an udder that is swollen, red, and hot to the touch. More often the first sign of mastitis is bad milk — milk that is lumpy, stringy, blood-tinged, or has a bad odor.

Goats should be tested frequently and routinely for mastitis using a test, such as the California Mastitis Test. Recent research suggests that, due to the chemical makeup of the goat's milk, the CMT may not be as useful in goats as it is in cows for indicating the presence of mastitis. However, it is still useful for eliminating mastitis as the cause of a problem with the milk.

Milking Your Goats

When you start your dairy goat operation, you will probably be milking by hand. Milking is not difficult, but it does require some practice. In a short time, you will become efficient and able to do it quickly. Each lactating goat must be milked twice daily, at roughly 12-hour intervals. Some dairy operators milk three times to encourage increased production, but twice daily is the norm.

Milking by hand

The following steps should be followed diligently and in the order presented:

Step 1: Prepare your equipment.

Place your clean milk bucket (start each separate milking with a clean pail), clean cloths, and a pail of warm soapy water in the milking area. Place your disinfectant and teat dip within easy reach. If you wish to test for problems, place your strip cup with the rest of your materials, so you do not forget to use it.

Step 2: Restrain your goat.

Before you milk a goat, restrain her in a stanchion or with a rope tie. A stanchion, also called a "head gate," or "milking stand," is a structure with a hole to restrain your goat's head while it munches on a snack, so its body is held still and steady for you to milk. It is always dangerous to leave a goat tied and unsupervised because the goat can quickly get tangled up, chew through the rope, or even strangle itself. However, a stanchion allows the milker to safely walk away from the goat for several minutes. Stanchions often have a raised platform on which the goat stands so the person milking does not have to bend over.

Training a goat to stand in the head gate is usually easy. Use a tasty treat or a little grain in a bucket to lure her head into the head holder and fasten her in. Reward her with the treat.

Step 3: Wash the udder.

Brush the doe before milking to remove any loose hair or dirt that might fall into the milk pail. Debris that lands in the milk, even though it will be filtered out, will affect the taste of your milk. Some milk buckets are made with a partial hood that keeps stray debris from falling in.

Using warm, soapy water and a clean cloth, wash any debris, manure, or dirt from the udder. After you have removed visible dirt, spray the udder and teats with a disinfectant. Dispense your disinfectant from a spray bottle, rather than using a communal container; it is too easy to pass bacteria between goats. You may also wish to wipe the flank clean along with the udder.

Step 4: Wash your hands thoroughly.

Wash your hands before milking each individual goat so bacteria are not transferred from one goat to another. Invest in a good hand lotion. Udder balm works wonders overnight on chapped hands. You can also use disposable gloves when milking.

Step 5: Use teat dip.

Dip each teat in your teat dip and wipe away excess with a disposable paper towel. Excess dip can taint the taste of the milk. The disinfecting properties take effect as soon as it touches skin, and it has served its purpose. Wipe it away religiously. It is also a good idea to put the teat dip in disposable 3-ounce paper cups and use a new cup for each doe.

Step 6: Milk your goat.

Many goat owners enjoy the process of milking by hand. It is peaceful, quiet time with your goat. Goats enjoy attention, and when accustomed to being handled, enjoy being milked. It is easier to learn with an experienced doe rather than one unaccustomed to milking. Always milk the goat on the same side. If your goat is accustomed to being milked from one side, it may take time for her to get used to being milked from the other. Provide some feed or the doe's grain ration for her to eat as you milk. Giving her something else to think about will increase her level of patience with you. Talking to her or singing to her will also calm her.

Place your milk pail under the udder. Begin by lightly grasping each teat in one hand and pulling down firmly, but *very slightly*. This allows the goat to "let down" her milk — getting the flow started. Next, squeeze each teat firmly without pulling it, in a steady, smooth downward motion. As each squeeze of your hand begins, first use pressure from your thumb and forefinger to prevent downward flowing milk from moving back up again into the canal. Keep that pressure steady while closing the third finger into the grip, then the fourth, and finally the entire hand. This is done in one smooth downward rhythm.

The first few times you milk a goat, your hands will cramp, and your back will hurt. Try to relax and breathe deeply. Be patient with yourself. Allow yourself some false starts. Countless human beings have successfully milked goats, sheep, and cows for 10,000 years. You can do it too.

The first few squirts from each teat should be discarded. (The resident barn cat is often nearby waiting with an open mouth to receive these.) Milk each teat until the udder is emptied and soft. Be careful not to over-strip the teat empty of milk, or you may damage it. Just stop milking when no more milk comes out of the teat during a normal squeeze.

Use a strip cup to check the milk for signs of mastitis or other problems. A strip cup is a metal cup with a permanent filter built into the top. Squirt the first squirt from each teat into the cup, and look for lumps, flakes, or any other abnormality that could indicate a possible problem. If you find something unusual in the strip cup, break out the CMT kit and check for mastitis.

Step 7: Redip the teats.

This is important to seal the ducts with disinfectant so that bacteria cannot travel up into the streak canal (passageway) and encourage mastitis. The teat opening is surrounded by sphincter muscles that relax to let the milk flow while you are milking. These remain relaxed and the opening loose for up to half an hour after you finish milking. During this period, the streak canal is most vulnerable to the entry of stray bacteria, and the final teat dip helps prevent this. Dispose of the dip; if you have another doe to milk, prepare new solution in a clean container. Never re-use the same dip because bacteria may be passed from doe to doe in contaminated teat dip.

Always begin the milking of each doe with clean materials: clean milk pail, clean soapy water, clean towels and cloths, clean dip, and clean hands. If you learn to think about preventing the spread of bacteria, you will have fewer veterinary bills and avoid many problems.

To begin, lightly grasp the teat in one hand and pull down firmly, but slightly. Squeeze each teat, without pulling it, firmly, and in one steady and downward motion.

When each squeeze of your hand begins, start with pressure from your thumb and forefinger, to keep the milk flowing downward, from going back up and into the canal.

Keep pressure stady, and close your thrid finger into your grip, next the forth, and finally the entire hand. Do this in a single smooth downward motion.

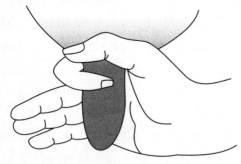

Diagram explaining how to milk a goat

Milking with machines

If you decide to increase the size of your herd to more than a few animals and you are the only person doing the milking, you will probably want to use milking machines. Machines are also useful if you have a physical limitation that makes milking difficult for you.

Milking machines are costly (almost $2,000 new for a machine that milks two goats), and unless you can increase your profit from milk and cheese sales, it may be difficult to justify the investment. It is possible to purchase refurbished and used milking machines for less. Manually operated vacuum pumps can also be used to mechanically milk goats, and these are inexpensive.

You will not be handling the udders of your goats as thoroughly and regularly as when you milked by hand, but you will still have to take time daily to examine udders for health problems and disease. Goats milked by machines can become unaccustomed to being handled by humans. Make a conscious effort to continue to put your hands on them daily, so they do not become difficult to manage over time.

Machine milking requires special stanchions set on platforms, on which goats stand for milking. The machinery itself must be disassembled and cleaned after each use. Udders must be cleaned and disinfected, just as with hand milking. The difference is that you will be able to milk several goats at once. You will spend less time milking and more time cleaning, disassembling, and reassembling.

When milking machines are used, the teat cups should be kept clean. Between goats, dip them into clean water and then a sanitizing solution. Teat cup liners should be free of cracks, milkstone (calcified milk), and ballooning. Machines should be properly cleaned, sanitized, and stored after each milking. Clean equipment will reduce chances of mastitis and lower bacteria counts in milk.

If you become a Grade A dairy, your equipment must be kept in top working condition, clean at all times, and always prepared for an unplanned government inspection.

Cleaning equipment

Whether you are milking by hand or machine, all equipment is cleaned several times daily. Pails or milk containers must be made of glass or of seamless stainless steel. Seams in metals and surface scratches in plastic and similar materials trap and hold microscopic bacteria. Bacteria cause spoilage and affect the taste of milk; they also can make your milk unsafe and spread disease and infection through your herd. Keeping containers scrupulously clean will help keep you, your goats, and your customers healthy, as well as keep your milk tasting great.

When you begin washing, soak containers in lukewarm water to discourage protein crust from forming on the surface. After soaking, wash in hot water with a disinfectant soap made for dairy use. Any scrubbing should be done with a stiff brush rather than a sponge or washcloth because both of these tend to hold bacteria and may shed minute fibers. At least once a week, soak containers and equipment in an acid-based dairy detergent, which will remove any mineral deposits on surfaces. Store your containers away from dust; inside a large cabinet is ideal. Rinse containers every time you take them out before you use them.

A clean barn will make it easier to keep your equipment clean. Keep floors swept and bedding fresh. Cement floors are sanitized with bleach and water. Add bleach to warm (not hot) water and slosh bucketfuls onto floors. Let it set for several minutes before mopping and rinsing off with a hose. Much of the harmful bacteria carried in feces, urine, and mud will be eliminated. If you are having a new cement floor poured, create a central gutter in the floor where rinse water can run and drain out of the barn. Make sure to allow floors to dry before putting down fresh bedding material because dampness under bedding can encourage mold, and damp bedding does not feel comfortable and warm to your goats.

TIP: Cleaning supplies

Poor sanitation quickly leads to the spread of infection and disease among animals and even between goats and people, and it can cause spoilage and contamination of your milk products. Sloppy sanitation practices can invite government scrutiny and fines and even lead to shutting down your operation. A clean farm means healthy goats, and healthy goats mean a quality product and money in your pocket.

Purchase these basic cleaning products for your operation:

- A large broom for sweeping barn and shed floors
- A large mop for wet use
- Buckets, towels, scrubbing brushes, and sponges
- Bleach for sanitizing
- Acid detergent
- A shovel for moving manure
- A rake for cleaning, or moving floor cover in stalls and pens

Processing milk

Milk is an ideal food; it is also an ideal environment for the growth of bacteria. As it approaches the teat opening, it encounters stray bacteria that have traveled from the opening up to the streak canal. When it leaves the opening and hits the air, it encounters dust particles laden with bacteria, and when it hits the bucket, encounters more. Good sanitation greatly reduces the amount of bacteria but cannot eliminate contamination altogether. It is important to create an artificial environment where the milk remains free of spoilage as soon as possible after milking.

Milk must be cooled to 40 degrees Fahrenheit (4.5 degrees Celsius) or below within an hour of milking. Refrigeration will not cool it fast enough, so most people immerse the containers of milk into buckets of ice water or into vats of continuously circulating cold water. If you must cool more than 6 gallons per milking, you may want to acquire a dairy cooler. A dairy cooler provides a steady supply of ice-cold water in which to place containers of fresh milk, which then cools the milk to the required temperature within minutes. In order to be certified as a Grade A dairy, you may be required to have a room designated exclusively for cooling and storing milk and a mandatory cooling machine.

Pasteurization

Milk that is not handled carefully and becomes contaminated can make anyone who consumes it sick. Pasteurization is the process of heating milk to a specific temperature and maintaining it for a specific length of time to kill harmful bacteria. In the 20th century, U.S. law began requiring that all milk sold for human consumption be pasteurized. However, many people believe that pasteurization makes milk less nutritious and detracts from its flavor and pasteurization is not necessary if the milk is handled properly.

The term "raw milk" refers to milk that has not been pasteurized. Many dairy goat owners keep their milk raw specifically because they appreciate the taste and nutritional value of raw milk, and they want it for themselves and their families. Advocates of raw milk point out that pasteurization gives milk a cooked flavor.

Milk was consumed raw for many centuries. People who routinely drink raw milk develop immunity to common bacterial infections. American tourists traveling in Europe sometimes become ill after eating yogurt, ice cream, or cheese because they do not have the same built-up resistance to dairy-borne bacteria as Europeans.

Milk for private consumption does not need to be pasteurized. If you sell goat's milk or an edible product made from your milk, such as yogurt, butter, or cheese for human consumption, you will need to pasteurize by law. If you run a large-scale operation, your milk will need to be transported to a local processing plant for processing that includes pasteurization within hours of your animals being milked.

Dairy farmers who know their own sanitation practices and the condition of their animals feel confident their milk is quite safe for their families to drink, as long as it is consumed while fresh (within a few days of milking). Enjoying raw milk is one of the benefits of dairy goat farming, but be aware of the danger so you can avoid potential milk-borne illnesses. Pasteurization keeps milk tasting fresh by destroying bacteria that proliferate and alter its flavor. Goat milk contains a unique enzyme called caproic acid that gives the milk a "goaty" flavor as it ages, usually after two or three days. If you choose to use your milk raw, both for safety and for the best quality and flavor, keep it chilled and drink it fresh.

When you transport milk, keep it in an airtight container that can be kept cool at all times, either by immersing it in ice water or by transporting it in a refrigerated vehicle. No sunlight from windows should fall on the container. If you are running a Grade A dairy, your local processing plant may provide safe transport of your milk to its facilities.

Off-flavors in goat's milk

There are many reasons why your milk may have an unpleasant taste. The majority of those causes can be controlled if you are aware of them. Many people blame an off-taste on the goat itself or assume it is caused by a buck in the vicinity, but that is often not the case.

In many eastern and northern European countries, goats are fed so their milk tastes "gamey." This is a desired trait in these countries, and goats that give milk with these strong flavors are highly valued as producers of milk and as breeders. Europeans often remark that goat's milk in America is bland. But if you are selling milk or edible milk products to Americans, you will want to keep goaty flavors out of your milk.

If you end up with a goat that gives strong milk consistently and persistently, you can try some folk remedies that may or may not be effective. One is a teaspoon of sodium bicarbonate in the feed daily — old-time farmers swear by this one. Another, favored by some modern goat farmers, is an occasional vitamin B-12 injection. If you wish to try the injections, speak to your veterinarian for advice about dosing because overdosing of any vitamin can lead to health problems.

People well versed in milk tasting use some common terms to describe off-flavors in milk. The table below shows some of these descriptive terms and their probable causes.

Causes of Off-flavors in Goat's Milk

Flavor	Cause
Barny	• Milking in a smelly barn • Not moving fresh milk from the barn quickly enough • Doe living continuously in a poorly ventilated barn even if she is milked in a place where air is fresh
Bitter/soapy	• The goat was consuming strong weeds before milking. • The goat was consuming strong feed before milking. • Biological makeup of milk in late lactation • Temperature of milk was altered too quickly, such as when warm milk is suddenly chilled.
Coarse, acidic/malty	• Unclean equipment • Failure to cool milk to 40 degrees Fahrenheit or below. • Milk cooled too slowly, which gave bacteria time to proliferate
Disinfectant	• Residue of chlorine bleach or other soap or disinfectant on equipment that was not thoroughly rinsed after cleaning
Feed/feedy/sweet	• Goats have consumed or have smelled plants, such as wild onions, garlic, silage, turnips, very green grass, ragweed, grape leaves, cabbage, honeysuckle, and others.
Foreign	• Goats have breathed fumes from paint, gasoline, or spray insecticides.
Metallic/oxidized	• Milk has made contact with corroded or rusted metal. • Containers made of copper, tin, or nickel • Goat's drinking water is high in iron or copper. • Milk exposed to sunlight
Musty	• Unclean hay or feed, unclean or stagnant water
Rancid	• Foaming resulting from vigorous milking in which too much air is mixing with the fresh milk • Sun penetrating a glass container for even a short period of time
Salty	• Mastitis
Utensil	• Goats drinking dirty water • Milk has made contact with unclean containers. • Milk has made contact with equipment that is not clean.

Some factors such as dietary changes and lactation cycles affect the composition and quality of goat's milk:

- Decreasing forage-to-feed concentrate ratio decreased milk fat and increased protein.

- Feeding of sodium bicarbonate buffer improved the percentages of fat and total solids.

- Later stages of lactation are associated with an increase in the content of fat, protein, and many minerals, and with a decrease in lactose, potassium, and citrate.

Chapter 7

Routine Care and Maintenance

airy goats must be moved and handled frequently, so it is important to understand their behavior. Goats cannot be easily herded like sheep; they tend to scatter in all directions. They can be led, and they will follow you if you establish yourself as the "herd queen" no matter what sex you are. Identify the doe leader of your herd and kindly make it clear that she must defer to you. Goats are highly motivated by food and will gladly follow a bucket of grain; be careful not to be overwhelmed by a crowd of eager, pushing goats.

You will need a lead rope with a latch that attaches to a halter and a halter suitable for your goats. Tack shops will have these items and will be able to help you get a halter that fits several goat sizes. Take the time once in a while to fasten a halter on a goat, whether you need to lead them or not. This is especially important with young goats, which need to get the experience of wearing a halter early in life so it is not frightening to them. An hour before you load the goats in the truck for that trip to the local school, fairgrounds, or veterinarian is not the time to get a goat used to the halter. So buckle one on, attach a lead, and walk the goat around with it.

Goats tend to avoid entering or walking across water, walking through areas of deep shade, passing through narrow spaces, or walking on slippery surfaces. Loud noises and sudden movements frighten them. They will gladly move from a dark enclosed space into the light and open air, uphill rather than downhill, and into the wind rather than away from it. A stubborn or frightened goat will lie down to resist being moved.

Goats can be trained using the same methods that work with horses and dogs: They respond well to being rewarded with food. As much as possible, lead a goat using a halter or a collar. Walk with your goat's shoulder at your hip, and have someone follow along to give her a gentle push from the rear if she stops.

Never use a choke chain or slip-style halter to tie up a goat. If the goat becomes frightened or agitated, it will frantically try to pull free and may strangle itself. Tie your lines with knots that can quickly be undone. Goats easily become entangled, and the more they are caught up, the more desperately they struggle to break free. Do not leave an untrained goat unattended when it is tied up, and keep an eye on tethered goats.

You can restrain a goat by cupping your hand under its chin and tilting its head up. Goats easily become stressed, and stress leads to health problems. Avoid harassing them, remain calm and patient, and give them time to understand what you want them to do.

Aggressive behavior includes glaring, raised hair along the top of the back, crowding, butting, biting, pushing, rushing, stomping of the front feet, and lowering of the head in a threatening gesture. If a high-ranking or aggressive goat exhibits this kind of behavior toward you, do not tolerate it. To thwart unwanted behavior, loudly say, "No!" and use a powerful squirt gun to squirt the goat in the face. To a goat, chasing is play behavior and being chased is a reward rather than a deterrent.

Hair and Skin

Trimming and brushing are important because anything that drops into the bucket during milking can affect the quality of the milk. Brush the goat before milking. Shaggy goats should have their hair clipped, particularly around the udder and flank. Trimming your goats in the spring will keep them cleaner and cooler during the summer and help discourage parasites.

During winter or when goats are kept in confinement, they should be brushed regularly with a stiff brush. When they are outdoors, goats scratch themselves by rubbing against fence posts and other objects to stimulate their skin and control parasites.

Trimming Hooves

Goats need their hooves trimmed two or three times a year to prevent infection and maintain integrity of the hoof. The frequency with which hooves need to be trimmed depends on the breed, time of year, and soil quality and moisture content. Trim hooves when you are performing other procedures such as worming or vaccinating. Do not trim hooves during periods of stress, such as extreme hot or cold weather, weaning, or late pregnancy.

Hoof trimming can be done with standard hoof shears, but you can also use horse hoof nippers, a hoof knife, and a rasp. Work in a well-lit area. If a goat's hooves are severely misshapen, the trimming may have to be done in several sessions.

1. Tie the goat securely or restrain her in a milking stand if the stand allows you access to all four hooves. It helps to have an assistant. Wear work gloves to protect your hands.

2. Squat beside the goat, sit on a low stool or stand with your back to the goat's rear end, and lean over. Lift the hind leg back and grasp hoof firmly. With a point of the tool or a stiff brush, clean dirt and manure out of the hoof.

3. Start trimming at the heel, working your way forward. Trim the heel even with the frog, the soft central portion of each toe, and then trim the walls level to match.

4. If the frog needs trimming, shave thin slices off with a knife, but stop as soon as you see a hit of pink. The finished hoof should be flat on the bottom and parallel to the coronary band (where the hoof and fur intersect).

5. As a final (optional) touch, you can smooth the hoof with a rasp.

To do the front hooves, squat beside a front leg and lift the hoof back, resting the foreleg on your knee while you work. Kids can be held on your lap to have their hooves trimmed.

Hooves are easier to cut when they are moist, such as when the goat has been walking in wet grass.

Avoid cutting into living tissue. If this does occur, stop the bleeding by applying pine tar or a bleeding powder. Bleeding will stop when the goat stands and puts pressure on the foot.

If one or more hooves are infected, trim the healthy ones first to avoid spreading bacteria. Disinfect your tools between each goat to prevent infection.

Chapter 8

Breeding Goats

airy goats must breed in order to produce milk. They start lactating when their kids are born and continue until they are well into their next pregnancy. A dairy goat farmer must learn to recognize when does are in heat and bring them together with a buck at just the right time so they can be bred. Then, the pregnant does must be managed and cared for for about 155 days until the kids are born and are either sold or kept to increase the size of the herd. Breeding goats requires a whole set of skills, including matchmaking, midwifery, and in some cases, nursing the newborn kids. Once the kids reach a certain size, you will either add them to your herd or sell them. An understanding of goat genetics is essential to optimizing the milk production of your dairy goat herd, even if you do not intend to keep any of your kids.

Staggering the Breeding Cycle

Goats have a breeding season during which they come into heat (estrus) every few weeks for a day or two. The breeding patterns of goats are related to the survival of their ancestors in the wild; wild goats bred during fall and early winter so their kids would be born in the spring when plenty of food was available. The Swiss breeds, which originated far from the equator, have their breeding season in winter when daylight hours are the shortest. Nubian and Pygmy goats, which have their origins closer to the equator, can breed year-round. Bucks have a breeding season similar to the does.

Does usually stop lactating and are dried off for the last two months of their pregnancies to give them time to rest before their kids are born, and they start lactating again. Some dry off sooner. That means each doe is producing milk for only 305 days out of the year, sometimes as few as 275. If all your does come into heat and are bred around the same time, they will also stop producing milk around the same time. This presents a challenge for dairy goat farmers, who ideally need a regular supply of goat's milk year-round. A family that relies on one or two goats to provide milk will have to do without for several weeks. The commercial demand for goat's milk is greater in the winter than in the summer, yet summer is when the goats reach their peak production. To keep the milk coming year-round, dairy goat farmers often try to stagger the breeding of their goats so some of them will always be lactating.

Does generally first come into heat in late August or September, when the days begin to shorten. If a doe is not bred during her first cycle, she will continue to go into heat at regular intervals until late in December. To ensure some of their goats will always be lactating, goat breeders try to stagger the breedings so some goats get pregnant in September and others at the end of December. This can be risky; if the doe is not pregnant by the end of her breeding cycles, she will not go into heat again for another nine months.

Some goat keepers try to influence the breeding cycle by controlling their does' exposure to light. A traditional method is to bring the does indoors in June and expose them to light for only seven hours a day to simulate the shortest days of winter, until they go into estrus.

When to Breed Your Does

Bucks and doelings (does that have not mated for the first time) reach sexual maturity at the age of 4 or 5 months, but sometimes bucks reach puberty earlier. For this reason, buck kids should be separated from does at the age of 2 ½ months. A normal young doe that has been properly fed can be bred successfully at 7 to 9 months, when she has reached 65 to 75 percent of her adult body weight (about 85 pounds). Breeding will help her udder to mature.

A doe should be prepared for breeding by flushing, increasing the energy in her diet to stimulate ovulation and conception. She should gain weight for two or three weeks before breeding and for about three weeks afterwards. This can be accomplished either by feeding her high quality forage or increasing her ration of grain. A doe bred during the first heat of her breeding cycle tends to have better milk production, and her kids will be old enough to breed by next season. You might want to postpone breeding until the second or third heat so the kid is not born during bitter cold winter weather or to stagger kidding so that your does are lactating at different times.

Recognizing Estrus in Your Does

Estrus is the state in which a goat's ovary contains a fertile egg, and her uterus is ready to establish it. A doe in estrus exhibits a number of external signs:

- Restlessness
- Constant bleating
- Attempts to mount other does
- Swelling and redness of the vulva
- Appearance of mucus on the vulva, which turns yellowish and cloudy toward the end of estrus

Estrus is easy to detect in a goat that has already kidded once but may be difficult to confirm in young doelings. Scratch the doe and press her at the base of the tail. If she stops and lifts her tail, she is probably in heat. A goat that is not in heat would tuck her tail under and dodge your hand.

If you own a buck, a doe in heat will be attracted and want to hang around the buck's pen. Bucks follow a similar breeding cycle to does. When a buck is ready to mate, he rubs himself with fluids from scent glands on its forehead and urinates on himself so that he gives off a strong odor. You can test whether a doe is in estrus by rubbing a cloth (known as a "buck rag") over the buck to absorb its scent. Put this cloth in a jar or sealed container. If the doe is in estrus, she will respond dramatically when you open the jar and let her smell the buck's aroma.

Estrus can be induced by the presence or smell of a buck or by exposure to the presence of other does in estrus. Warm temperatures sometimes inhibit estrus while cold temperatures stimulate it.

Estrus lasts for about 36 hours. The ovary does not release the egg until just near the end of the outward signs of estrus, so the best time for mating is about one day after estrus begins. It is safe to take the doe to the buck as soon as you notice she is in heat; if possible, breed her again 12 hours later to ensure conception.

Mating

Once your doe is in estrus, she must be brought together with the buck within 24 hours. Many dairy goat owners do not keep a buck but instead pay a fee for stud service. You can also lease a buck from a local breeder and keep him with your does for several days. If you do not observe the breeding, the buck should remain with the does for at least three weeks, a complete breeding cycle. Occasionally, a breeder will accommodate your does to stay with the buck for the breeding period.

Stud service is carried out in several ways. You can transport your doe to the buck and drive her home again after a brief encounter, arrange to board her at the farm with the buck for a few days, or have the buck brought to your farm (this is called driveway service). Make arrangements with the buck's owner well in advance to be sure the buck is available when you need him during the busy breeding season. If you plan to register the offspring, be sure to ask for a service memo, a document signed by the buck owner verifying that the buck bred the doe and the date. Most breeders will give a second breeding free if the doe does not get pregnant the first time.

When a doe in heat is brought to a buck, he greets her and begins smelling her from all sides. If she urinates, he smells or tastes it and makes a grimace, called a flehmen, holding his head high and drawing up his upper lip. Then he strikes his forefeet on the ground, lays them on top of the doe, and attempts to mount her. If she is at the peak of estrus, she tolerates the mounting and co-operates by lifting her tail. The copulation is complete when the buck raises himself high with a powerful thrust and throws his head back. If the doe is still in heat 24 hours after the first mating, a second mating should take place. A healthy mature buck can mate about ten to 20 times a day.

Choosing a Buck

Some goat keepers breed their does solely to stimulate lactation (freshening) of does in their dairy herd and quickly sell the kids. Others breed their does with an eye to improving their herd with offspring that produce more milk, are stronger and healthier, and are productive longer. They select the best kids from each generation and sell their inferior goats. Whether you are breeding to freshen your does or to increase your herd, always seek out the best possible buck to impregnate your does. You may be tempted to use a buck from a neighboring farm just because it is convenient, but remember that the buck's genes influence not only the quality of its offspring, but the health of the mother and kid during pregnancy. A buck could carry a genetic flaw that results in abnormal fetuses or miscarriages. By selecting high-quality bucks, you will also produce superior kids that can be sold at higher prices to other dairy farmers.

You are keeping dairy goats because you want milk. Every year you make a considerable financial investment in feed, medications, and veterinary fees for your goats, to say nothing of all the time and effort you put into their care. Naturally you want the maximum return on your investment — as much milk as possible and a healthy herd made up of goats that continue to be productive for eight or nine years or longer. You want to find a buck whose offspring are high-yield milkers with firm, well-supported udders.

You can find a registered buck for breeding in your area by looking online at goat registries for various breed associations or searching the American Dairy Goat Association (ADGA) genetics website (**www.adgagenetics.org**). Your agricultural extension office or veterinarian may be able to tell you about local goat breeders who offer stud service.

When considering a buck, first look at his pedigree, if he has one, and the milk production records of his forebears. An undocumented buck or a grade can produce wonderful offspring, but a pedigree and official records are a sort of insurance that the buck will produce good offspring.

Next, look at the physical characteristics of your does and of the buck and his forebears. If possible, examine the buck's mother and granddam. If your does have a weakness such as poor udder attachment, the genetic background of the buck should counteract that flaw. Look for a buck descended from a family of does with strong, well-attached udders. To improve your herd, always breed your does with a buck from a family of even better does. That way the offspring will be better milk producers than their mothers.

Artificial Insemination

In artificial insemination (AI), frozen sperm from a donor buck is thawed and inserted into a doe in heat. If you are aiming to improve your herd, artificial insemination allows you to select from a wide range of bucks with documented backgrounds. You can select a different buck for each of your does. The procedure can be performed by a veterinarian, another goat breeder who has experience, or you can learn to do it yourself. Defrosted semen is viable for only a few hours and should be inserted just at the end of the doe's heat cycle. Success depends on knowing exactly when a doe went into heat and how long her normal heat cycle is.

A "straw" of frozen semen costs as little as $5 to $25, but the equipment for storing the frozen semen requires an initial investment. Frozen semen is stored in a liquid nitrogen tank at -320 degrees Fahrenheit. A used liquid nitrogen tank costs $100 to $300 ($600 new), and it costs about $35 to $50 to refill the nitrogen tank every two months or so because the nitrogen evaporates. A single tank can hold 300 to 1,000 straws, so it is possible to share a tank with other goat keepers.

TIP: Equipment and supplies for artificial insemination

Liquid nitrogen tank

Straws of frozen semen

Tweezers for removing straws from the tank

Straw cutter for cutting off the wax plug keeping the semen in the straw

Open-ended glass speculum

A small light to make the cervix visible

Thermos of warm water to defrost the semen

Thermometer for measuring the temperature of the water

Insemination gun (a thin metal tube with a plunger to push the semen out of the straw)

Disposable plastic sheaths to hold the straw while in the insemination gun

You can purchase frozen semen online from suppliers such as Superior Semen Works (**www.superiorsemenworks.com**) and BIO-Genics, Ltd. (**www.biogenicsltd.com/index.html**) and sometimes directly from a buck's owner.

The advantages of AI are that you have access to a wide selection of bucks instead of being limited to those within driving distance of your farm, and you do not have to worry about transporting goats or possibly carrying disease from one farm to another. Sires are screened for sexually transmitted diseases before semen is collected.

Conception rates with AI vary from 50 to 70 percent and are affected by a number of circumstances, such as the sensitivity of the goat keeper to the goat's estrus and the amount of stress the doe experiences during the procedure. Does should be handled gently and reassured. Sometimes being transported to a veterinarian can upset the doe; if you have several does in estrus at the same time, you may be able to have the AI practitioner come to your farm.

If you use AI, be sure to obtain a Record of Artificial Insemination stating the doe, the buck, the source of the semen, and the date of insemination.

Keeping a Buck

It might seem logical for a goat keeper who wants to be self-sufficient to keep a buck to service the does. Whether this is feasible depends on a number of factors, including the size of your property and facilities, the size of your herd, and whether you will be able to earn some income by offering stud service or selling superior kids to other goat keepers.

It costs as much to feed and care for a buck as for a doe; if you are keeping only two or three does, keeping a buck will essentially double the cost of your milk. A buck is often kept together with a herd of meat goats because natural mating behavior results in a higher rate of pregnancy. Bucks cannot be kept together with a herd of dairy goats when they are in late pregnancy or after the kids are born. The presence of a buck in the herd can cause a strong "goaty" flavor in their milk due to the extra hormones in the herd. A buck can be aggressive toward kids and may impregnate the does again too soon after the births. The buck needs a separate pen and enclosure, preferably at least 50 yards away from the does. Bucks are particularly strong and aggressive, so this pen and enclosure must be exceptionally well built and reinforced to prevent the buck from escaping and rejoining the ladies. Because goats prefer company, you will probably need to keep another buck or a wether, a castrated male goat, as a "friend" for the buck.

A buck kid is cute and playful, but when he matures, he becomes aggressive and dominating and may be difficult to handle, especially for a child. During breeding season, bucks cover themselves with scent and urine and give off a powerful aroma that seems to permeate everything. Neighbors may find this odor offensive, and many residential areas that allow the keeping of female goats prohibit the ownership of bucks.

After the first breeding, you will be mating your buck with his daughters. Sometimes this kind of inbreeding strengthens desirable qualities such as milk production and udder strength in the offspring, but it can also exaggerate flaws and weaknesses. Most bucks are sold as kids before they have been bred to show the kind of offspring they produce. An older buck that has demonstrated its superiority is expensive if it is still in its prime. A buck that is a good mate for one of your does might produce kids with undesirable traits from another of your does. This is not a problem if you are only using the buck to "freshen" your does and selling all the kids soon after birth. If your goal is to increase or improve your herd, after one or two generations you will need genetic input from another buck.

Chapter 9

Pregnancy and Kidding

ost does experience normal pregnancies. Depending on the age, breed, and previous pregnancies, 145 to 155 days after the last breeding date your doe will deliver one to five kids. Multiple births usually occur a little earlier; an older or poorly nourished doe may give birth later.

It can be difficult to tell whether your doe is pregnant during the first three months. The first sign of pregnancy is that the doe does not go back into heat after her last cycle. Watch the doe closely three weeks after she has been bred. A diagnostic lab, a veterinarian, or a mail-in service can do pregnancy tests. Typically, a sample of urine or milk is tested for the presence of estrone sulfate, a hormone produced by a living fetus about 35 days after conception. If the doe is not pregnant and you are still in the breeding season, it may be possible to breed her again. You can continue milking a doe that is not pregnant right through to the next breeding season, though her milk production will not be as high as a goat that has given birth.

Later in pregnancy, you will be able to feel the presence of the kids through the goat's abdomen. Some veterinarians perform ultrasounds on livestock, but this is not necessary unless you are overly concerned about a doe.

Miscarriage

A goat that was bred and then comes back into heat may have miscarried a fetus. Miscarriage is not uncommon among goats. There are a number of causes: disease, poor nutrition, poisoning, death of the fetus, or the natural rejection of an abnormal fetus. Diseases that cause miscarriage include leptospirosis and vibriosis; your veterinarian or local agricultural extension office can tell you if goats in your region need to be vaccinated against these. Injury is a common cause of miscarriage. Injury can result from does butting each other to establish dominance or by climbing or jumping off elevated objects. A pregnant goat that is new to your herd, or one that is overactive, should probably be put in a private stall where she will be more sedate.

You may not always be able to detect miscarriage in the early stages of pregnancy. Signs of miscarriage include bloody discharge and changes in behavior such as loss of appetite, a dazed appearance, or reluctance to mingle with the herd. If you suspect a miscarriage has occurred, watch for the expulsion of the placenta. Do not rebreed a doe that has suffered a miscarriage until the next season.

It is important to try to determine the cause of miscarriage. A live or recently dead miscarried fetus indicates a miscarriage caused by some kind of stress. A miscarried fetus that is decomposing or appears to have been dead for some time might indicate that the doe is a habitual miscarrier and should be culled from your dairy herd.

Record dates and causes of miscarriage on the individual health record that you keep for each doe. If several goats in your herd miscarry or give premature birth, they may be infected with chlamydia or another disease, and you should consult your veterinarian.

Caring for a Pregnant Doe

Milk-producing (lactating) goats should be dried off two to three months before the kids are due to allow their bodies to nourish their kids and build up reserves. Five or six weeks before the kids are expected, boost your goats' C/DT vaccinations, trim their hoofs, and worm them if necessary. *See Chapter 10 on goat health for more information.* Read labels on worming preparations carefully, or consult your veterinarian, because some can cause miscarriage.

Around this time, begin supplementing your does' diet with concentrate or high-quality forage. Consult your agricultural extension office or a local goat keeper for advice on local conditions. In areas where soils are selenium-deficient, pregnant does should receive a Bo-SE (selenium/vitamin E) injection. Ensure that your goats get enough exercise. Pregnant dairy goats should be well nourished but not fat. *Problems of Pregnancy in Chapter 10 will address these issues in more detail.* Excess fat can lead to serious complications before and after giving birth.

Unborn kids accumulate 70 percent of their birth weight in the last six weeks of pregnancy; during this period, your goat should visibly fill out. To accustom your doe to the milking routine she will follow after kidding, occasionally place her in the milking stand and feed her a little grain. While she is in the stand, you can look her over and inspect the udder for swelling or other problems.

About ten days before you expect your first kid, assemble your kidding supplies. If your goats will kid indoors, clean and disinfect the pens where does and kids will be kept, or construct new ones in a draft-free but well-ventilated area. Allow 25 to 35 square feet of space for each doe and her kids. Litter the floor with dust-free bedding (sawdust is not good bedding for newborn kids because it can trigger respiratory problems) and arrange for watering and feeding. Remember that a tiny kid can drown in a standing bucket of water, so use small or elevated watering containers.

A week before the kids are due, clip the hair around the does' udders and escutcheons (the area between the udder and tail), vulvas, and tails.

Drying Off

Drying off is the process of stopping lactation in a dairy goat. In late pregnancy, the milk production of many does stops naturally. If this does not occur two to three months before the kids are due, stop milking to allow the goat's body to build up reserves for lactation after the birth. Some goat keepers recommend drying off gradually by reducing milking to once a day and then every other day. Other farmers believe it is better to stop milking the doe altogether and let the pressure in her udder naturally put a stop to milk production. When milking ceases, the udder forms a mucus plug to prevent bacteria from entering through the teat. Each time you milk the goat, this plug must be re-formed, which increases the chances of infection, and the goat's udder receives another signal to produce milk. Gradual cessation of milking can result in fibrosis of the udder and lower milk production in the future.

To dry off your goat:

- Stop milking. After the last milking, some goat keepers recommend using a dry-cow antibiotic infusion to reduce the possibility of mastitis. This is done by injecting antibiotics into the teat canal. Use a teat dip.

- Stop feeding the goat high-nutrient foods that encourage milk production. Switch to a dry doe ration with lower concentrate and protein content.

- Continue dipping the teats twice daily for the next four to five days.

- If substantial pressure still exists in the udder after four days, milk out the doe, give another antibiotic infusion, and start over.

Just before she gives birth, a doe's udder will begin to "bag up" with milk in preparation for her kids. A high-yield milker may need to be milked once or twice at this time if her udder becomes too full.

Signs a Doe is Ready to Give Birth

A doe almost ready to give birth becomes nervous and restless. She may lose her appetite; turn around, lie down, and get up again repeatedly; bleat; and turn her head towards her tail. The vulva becomes red and swollen and may trail a long string of clear mucus. The mucus becomes opaque and yellow when the birth is imminent. The goat may withdraw from the herd, sometimes taking another doe along with her. Other does in the herd will show exaggerated interest in a doe that is about to kid. If they come too close, she will reject them. Do not isolate a doe from the herd immediately before she gives birth as this can upset her; wait until after the kids are born to move her to her own stall.

Near the end of pregnancy, a doe's ligaments loosen and stretch to prepare for the birth. The tailbone becomes elevated, and a hollow appears on either side of the tail. As birth approaches, feel for the goat's ligaments twice a day by grasping the goat just above the tail. In a doe that is not pregnant, these ligaments are firm and tight. As pregnancy advances, the ligaments feel loose and elastic. About 12 hours before birth, they seem to vanish entirely.

A veteran doe usually begins to bag up — her udder fills with milk — about ten days before birth. Some does, however, do not bag up until later. A doe experiencing her first pregnancy will begin to build an udder four to six weeks before birth.

About 140 days after the doe was bred, start watching for signs that she is about to give birth. Make a note of pre-kidding signals on the goat's health record. No two does behave in exactly the same way, but a doe is likely to repeat the same kidding signals every time she gives birth. If you take good notes the first time, you will be better prepared the next year. You should be able to feel the kids on the right side of the doe until about 12 hours before birth. Then the uterus begins to tense and one of the kids will be forced into the neck of the womb. At this point, the slope of the doe's rump becomes more horizontal, and birth can be expected in about two hours.

Birth

Most births take place naturally without any assistance. It is better not to interfere unless absolutely necessary; just observe the birth calmly and be ready to step in if there is an emergency.

A goat can give birth standing and lying down. The first thing you will see is a translucent bubble, the water sac, which can burst at any time. You will see one hoof and then another, and then a little nose, wrapped in the amniotic sac. Eventually this sac breaks. The birth slows, giving the birth canal time to dilate a little more for the head to pass. Once the shoulders are delivered, the rest of the kid plops out. Most kids are born with the front feet forward and the head lying on them; less often the hind feet come out first.

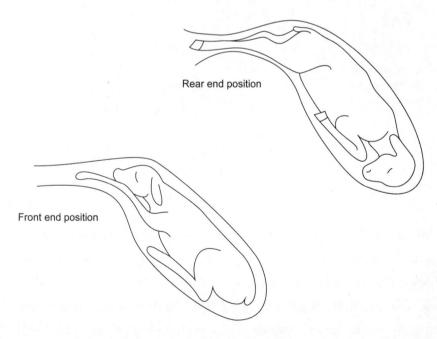

Positions a kid can take during birth

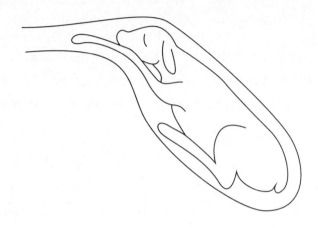

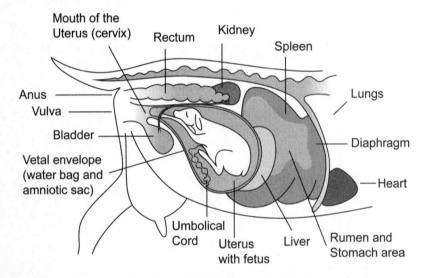

Position of kid during normal birth

The kid lies motionless for a short while, then lifts its head and tears the amniotic sac. Except in a case where the kid is frail, it does not need help removing the amniotic sac. The umbilical cord will also break by itself. Disinfect the end of the umbilical cord by dipping it in a small container, such as a shot glass, containing 7 percent iodine for several seconds. Use a fresh container of iodine for each kid. Do not omit this step. An umbilical cord longer than

2 inches needs to be trimmed. If the umbilical cord continues to bleed, use a navel clamp about 1 inch below the kid's belly, or tie it off with clean dental floss.

The mother immediately begins licking the kid to remove mucus and stimulate its breathing. If a kid is frail and weak, you can draw off the mucus and rub gently with a towel to stimulate breathing, then place it in front of its mother to be licked. If there is more than one kid, the mother will stop licking to turn her attention to the next delivery. Move the first kid to the side so it will not be stepped on. Do not leave the doe and kids unattended until you are sure all of them have been born. A doe can have from one to as many as five kids; multiple births are more common in some breeds.

Soon the kid will try to stand up. It will begin to suckle about half an hour after birth. Milk one stream from each teat to remove the mucus plug. If the udder is swollen, the kid may need some assistance with the teat. The first milk, known as colostrum, is important for the kid. It contains antibodies that disappear from the doe's milk after the first 24 hours. A newborn kid should ingest its first meal of colostrum within half an hour to two hours after birth. If the kid does not nurse, milk the doe and bottle-feed it colostrum.

TIP: Using a stomach tube

A kid that is too weak to nurse must be fed with a stomach tube. A stomach tube is a length of small, flexible plastic tubing such as that used for catheters. You can purchase stomach tubes from a goat supplier. The tube is attached to a 60 ml syringe.

Slowly and gently push the tube down the kid's throat. The kid will often swallow as you push the tube down.

When the tube has reached the kid's stomach, attach the syringe and depress the plunger slowly to deliver milk or colostrum directly into the kid's stomach.

Make sure the tube is in the kid's stomach before beginning to push the plunger; otherwise, you might force fluid into the lungs.

The kid's suckling will assist the mother with expulsion of the afterbirth. Some does eat the afterbirth; this is completely natural. If the doe does not eat it, dispose of the afterbirth as soon as possible. If the afterbirth is not expelled within a few hours after delivery, call your veterinarian. If the placenta is visible but not fully expelled, do not attempt to pull it out because you could cause the goat to hemorrhage.

Most births do not require assistance. If a large kid gets stuck passing through its mother's hips, give a gentle tug on the front legs. A kid born hind legs first (the bottoms of the hoofs are turned upwards) should be helped with a gentle tug because it could drown in amniotic fluid if it remains too long in this position. If a kid presents itself in any other position, you can try to reposition it. Wash your hands and wash and dry the doe's vulva with a mild dishwashing soap. Lubricate your hand with Vaseline® or dishwashing soap and insert it gently into the vagina between contractions to press the kid back into the uterus and manipulate it into a correct birth position. As you move a tiny hoof into the birth canal, cup your hand over it to protect the uterus from being torn or injured. This should be done by someone with knows what he or she is doing; call your veterinarian or an experienced local goat keeper. If the goat has been in hard labor for 45 minutes and you have been trying unsuccessfully to help her for 15 minutes, it is time to call for help.

After the birth, offer the doe a bucket of lukewarm water with a little molasses or cider vinegar in it and a feed of hay if she wants it. Move the kids into a draft-free box or pen where they will be reasonably warm. Remove any wet bedding from around the doe. For the next two weeks, she may continue to have a slight discharge. If the weather is warm, she may shed quantities of hair. Brush her regularly. Over the two weeks following the birth, gradually increase the doe's grain ration until she is back on her regular milking diet.

TIP: Colostrum

Colostrum, the thick yellow substance secreted by a doe's udder after birth, is essential to a newborn's well-being. It contains antibodies that strengthen the kid's immune system against disease and helps to clear and condition the newborn's digestive system. Does produce colostrum for one to four days after the birth. Kids should be fed exclusively on colostrum for the first two days of life. If you are hand-feeding your kids, milk the colostrum, and give it to them in bottles.

The viruses that cause Caprine Arthritis Encephalitis Syndrome (CAE) and Johne's disease pass from mother to newborn in the doe's colostrum. *See Chapter 10 for more information on these viruses.* Heating the colostrum to 131 degrees F (55 degrees C) for one hour destroys these viruses. Do not heat it above 140 degrees F (60 degrees C) because all the antibodies will also be destroyed. To prevent the newborns from acquiring these diseases, you must make sure they do not nurse directly from their mothers. Bottle-feed them colostrum that has been heated and then cooled to body temperature. Extra colostrum can be fed back to the mother or frozen for future use.

To avoid scorching the colostrum, heat it in a double boiler or slow cooker for an hour, or heat it to 135 degrees F (57.2 degrees C) and keep it in a thermos for one hour. If a newborn kid nurses from its infected mother even once, all of your efforts will have been wasted. When you are not able to be present at the birth, buy special tape from a goat supplier to seal the goat's teats so the newborn cannot nurse.

If you cannot obtain colostrum from the mother, you may be able to get frozen colostrum from another goat breeder. If fresh or frozen goat colostrum is not available, you can substitute cow or sheep colostrum. You can purchase colostrum replacements from a goat supplier.

Keeping Kids Warm

Occasionally, a kid is born unexpectedly when you are not there to supervise. You may arrive at the barn to find a cold, shivering newborn. If it seems to otherwise healthy, do not bring it into your house to warm up. Place it in a draft-free box or pen lined with a blanket, and if necessary, use a heat lamp to warm it up. Do not let it get too hot; overheating is just as harmful as being too cold. Kids are generally comfortable at temperatures 50 degrees F (10 degrees C) or higher.

A kid that is severely chilled, still wet, and almost lifeless can be revived by placing it up to its nose in a bath of water at 104 degrees F, the temperature of the womb it just exited. When it has recovered, dry it well, wrap it in a blanket or towel, place it in a sheltered place, and watch it closely. A newborn that appears to be very cold can also be suffering from low blood sugar (hypoglycemia) — shivering, ruffling its fur, and arching its back. Warm the kid and use a stomach tube (see below) to administer at least 25 mg of 5-percent glucose solution. When it appears to be reviving, feed it 2 ounces of colostrum, using the stomach tube again if necessary. As soon as it is active, take it back to the barn.

If you end up nursing a weak kid in your warm kitchen, it will probably stay there for the rest of the winter because it will not be able to adapt to the cold easily when you return it outside.

What to Do With the Kids

If you kept all the kids born to your does every year, you would soon be overrun by goats. Roughly half of all kids are bucks; because one buck can impregnate 50 to 100 does in a season, there is no need to keep more than one or two adult bucks for your herd. You might want to raise a few of your best doelings to increase your herd or to sell as dairy stock to other farmers or raise your unwanted kids for meat. Inspect your kids after birth. Look for extra teats on the doelings. Double teats or extra teats that are too close to the real teats can interfere with milking. Bucks that have extra teats should not be used for breeding because they could pass this trait to their offspring. A doeling that has a small pea-like growth on its vulva is a hermaphrodite and will be infertile; avoid breeding her mother with the same buck again.

Raising kids is a lot of work and requires extra space and special facilities. Kids are usually not left with their mothers in a dairy goat herd because the priority is milk production. If the kid is left to nurse whenever it wants all day long, there may be no milk for you at milking time, and there is no way to measure the doe's milk production or even to know if the kid is getting enough nourishment. The doe needs to become accustomed to the milking routine, and the kid must become accustomed to being handled. Kids that remain with their mothers may adamantly resist separation later on and may continue to nurse long after they should have been weaned. Some goat keepers believe that nursing ruins a milk goat's teats. For this reason, kids in a dairy herd are kept with other kids of the same approximate age and fed from pans or bottles.

There are some cases in which kids should be allowed to nurse. If a doe has a tight or congested udder after giving birth, bring her kid(s) to nurse every few hours for several days to relieve the condition. Nursing also stretches and enlarges the teats of a first-year doe if they are too small for the hands of the person milking them.

Raising Kids

Kids, unlike sheep, do not maintain a lasting relationship with their mothers. If left to grow up in the herd, they spend much of their time with other kids, returning to their mothers only to suckle, and becoming independent when the milk dries up. After they have had their early colostrum, dairy kids can be taken from their mothers and raised together in a separate nursery area. They will sleep a lot during the first two weeks of life, then become increasingly active. Change their litter often, and give them plenty of opportunities to exercise. Kids love to play on an upturned box, a platform, or a barrel laid on its side.

The nutrition and care kids receive during their first year determines how healthy and productive they will be later in life. Kids need to spend as much time as possible outdoors. Because their young bodies are susceptible to worms and parasites, they should be separated from adult goats in a pasture that has not been grazed for six months or more. If this is not possible, a paved run that can easily be cleaned is the next best alternative. Many goat keepers use portable hutches as kid shelters that can be moved to new pasture for each new group of kids.

Keep a health record for each kid, beginning with its birth weight, height at the withers, and heart girth. Weigh the kids every two weeks. An easy way to weigh a kid is to step on a bathroom scale without the kid, then step on it again holding the kid, and subtract the first weight from the second.

Spend at least ten minutes a day with your kids and handle them often so they get used to cooperating with people. Talk to them and use their names so they learn to come when called. Put a collar on each one and use it to move the kid where you want it to go. When you need to inspect a kid or perform a procedure, sit down and hold the kid on your lap or between your legs.

Feeding Kids

Goat keepers swear by a whole range of feeding programs and schedules, most of which seem to work well. The most important thing is to be consistent and stick to the same method and feeding schedule every day. Most kids begin eating their first hay after one week and can be weaned from milk entirely at eight weeks.

Bottle or pan feeding

Kids can be fed milk using either pans or bottles. Once a kid has learned one method, it is difficult to get it accustomed to the other. Pans are easier to wash, sterilize, and fill than bottles, but many goat breeders do not recommend pan feeding because the kid must lower its head to drink, and this may allow milk to get into its rumen. When a kid nurses from its mother, its neck is raised at an angle that closes off the passage to the rumen and allows the milk to go straight into its abomasum. Kids also tend to step in the pans or knock them over in the excitement of feeding. To accustom a kid to pan feeding, let it suckle on your finger a little, then lead it to the pan and gently hold its nose in the milk until it starts to drink. You may have to repeat this process several times. Always wash and sterilize the pans after each feeding.

Many types of bottles and nipples are used for bottle-feeding. The simplest (and cheapest) is a nipple that fits over an ordinary soda bottle. You can use a nipple designed for feeding lambs or an ordinary human baby nipple with a large X cut into the end to enlarge the opening. Bottles and nipples especially designed for feeding goats can be purchased from goat supply catalogues for $6 to $8. Once you have taught a kid to drink from the bottle, the bottles can be placed in a holder or rack placed just high enough that the kid has to raise its head to drink. Raise the height as the kid grows. If you are feeding multiple kids, you can also use a commercial feeding device, usually referred to as a lamb bar — a large container with several nipples sticking out around its sides. Each nipple is connected to a tube that goes down to the bottom of the container so the kids suck up all the milk as they drink. You can also purchase the nipples and tubes and make your own device out of a bucket.

Teaching a newborn kid to drink from a bottle may be a challenge because it does not know what a bottle is and wants to drink from its mother. Nipples should be warm and soft, and the milk should be warmed to around 104 degrees F (39 degrees C), the normal body temperature of a goat. Warm the milk to 104 degrees F so that it cools to no less than 98.6 degrees F (37 degrees C) as you are feeding. Hold the kid in your lap and gently but persistently dribble milk from the nipple into its mouth until hunger takes over. After the first few days, the milk can be fed warm or cold, but most goat keepers continue to feed warm milk. The important thing is to be consistent.

What milk to use

Fresh or frozen goat's milk is the easiest and most natural milk to give a kid. If this is not available, you can use fresh raw cow's milk, or regular whole cow's milk from the grocery store. Do not use canned goat's milk. You can add 3 tablespoons of corn syrup, which adds nourishment for the kid, to each gallon of whole milk or make up one of these mixtures:

- Five parts whole milk to one part dairy half-and-half

- One cup of buttermilk and one 12-ounce can of evaporated milk to every gallon of whole milk

If you are planning to sell your kids or do not want to buy cow's milk, you can use a powdered milk replacer, available from feed stores or dairy suppliers. Use only a milk replacer formulated specifically for kids and follow the instructions on the label carefully each time you mix up a batch. Do not use soy-based milk replacers or those intended for young of other species. Mix up only enough milk replacer for one day at a time, and do not pour what is left over after feeding back into the container. Do not abruptly switch brands or products. Powdered milk replacers can cause diarrhea, bloat, digestive problems, and floppy kid syndrome if used improperly. *Floppy kid syndrome is discussed later in this chapter.*

Feeding Schedule

Newborn kids should be fed at regular intervals every six or eight hours for the first three days, every eight hours for the first week, and two or three times a day afterwards. Some goat keepers feed only twice a day after the first week. The feeding schedule should be consistent every day. If you must be absent at a regular feeding time, you can leave a feeding bucket where the kids can access it, cooled with ice or mixed with a buttermilk culture to keep it from spoiling.

After one week, start giving the kids fresh, leafy hay in an easily accessible manger or feeding rack so they can start developing their rumens. Cut back a little on the amount of milk, so they will feel hungry enough to be interested in the hay. After three weeks, start offering the kids small amounts of grain; if you are pan feeding, sprinkle the grain in their milk.

Suggested Feeding Schedule for Kids

Age	Feed	Amount per day	Frequency
Birth–3 days	Colostrum	12–16 oz.	½ cup, 4 – 5 times per day
4–7 days	Goat's milk	12–24 oz.	1 cup, 3 – 4 times per day
1 week	Milk	36 oz.	1 cup, 2 – 4 times per day
2 weeks	Milk	32 oz.	1 ½ – 2 cups, 3 times per day

Age	Feed	Amount per day	Frequency
3–8 weeks	Good hay	Freely available	
	Water	Freely available	
	Milk	32 oz.	2 times per day
	Good hay	Freely available	
	18% starter grain	As much as the kid can eat in 15 minutes	2 times per day
	Water	Freely available	

A good starter grain mixture for kids is:

- 40 percent oats and barley
- 20 percent wheat bran
- 12 percent groats (soy chips)
- 25 percent linseed oil cake meal
- 3 percent mineral mixture

Grain should be coarsely ground, and you can also supplement with a tiny amount of limestone (calcium carbonate) and vitamins A, D, and E.

After eight weeks, gradually start decreasing the amount of milk. Kids weigh about 7.5 pounds at birth and gain around 10 pounds per month. They should be weaned when they reach about 40 pounds, which is typically at 3 months of age. After that, feed kids approximately 1 pound of grain per day. The amount of grain a kid needs varies depending on the quality of the hay it is receiving and how much weight it is gaining. Cut back on grain if the kid is gaining weight too rapidly or if you cannot feel its ribs through the skin. By the time a doe is 6 months old, it can be put on a milking ration, and at 7 months, it can be bred for the first time.

Disbudding and Castration

Both male and female goats have horns; a goat that is born horn-less (polled) is often infertile. Most dairy goats have their horns removed because they cannot easily fit their heads into feeding racks and milk stands and because the horns can cause injury to their human handlers or to other goats. This is usually done by disbudding, cauterizing a kid's horn buds, when it is 3 to 14 days old. If a kid has tight skin and curly hair over its horn buds, it has still horns and must be disbudded.

An electric disbudding iron with a ¾- to 1-inch tip can be purchased from a goat supplier. Here are the steps you should follow to disbud a goat:

1. Heat the iron until it is hot enough to leave a burn mark on wood.

2. Hold the kid on your lap, or place it in a restraining box, and trim the hair around the horn button with a small pair of scissors.

3. Grasp the kid by the muzzle and press the hot iron on the horn button for a count of 15. The kid will scream and struggle, and you will smell burning hair. There should be a copper-colored ring around the horn bud.

4. Comfort the kid and allow the iron to heat up again. The do the other horn bud.

5. Spray the area with an antiseptic spray, taking care not to get the spray in the kid's eyes.

6. Give the kid a bottle of milk to console it.

Some goat keepers give the kid a painkiller such as aspirin 30 minutes before disbudding. Disbudding can be done without an anesthetic until the kid is 1 month old. After that, it is better to let a veterinarian do the job. A local goat breeder might disbud your goats for a small fee. If the horn bud is not completely removed or thin, misshapen horns called scurs start to grow, you will have to repeat the disbudding.

Disbudding can also be done using a caustic dehorning paste something like a wart remover, available from farm supply stores. Clip the hair around the horn buttons, then cover each horn button with a disk of adhesive tape. Cover the surrounding area with Vaseline to protect the skin. Remove the adhesive tape and apply caustic paste to the horn buds. Hold the kid still for half an hour to prevent it from licking the caustic or smearing it on something. Be careful that the kid does not rub the caustic paste on itself or on another goat. It can cause burns and blindness.

If you are working alone or have several young kids to disbud, you can immobilize a kid inside a burlap bag with a corner cut off for the head to stick through, or a in a specially built wooden disbudding box that confines the kid and holds the head in place. Goat suppliers sell these boxes, or you can purchase the head restraint and build one yourself.

Castration

Bucks that are going to be slaughtered before they reach maturity do not need to be castrated. A buck that is going to be kept as a wether should be castrated as soon as its testicles descend into the scrotum, when it is between 1 week and 3 weeks old. Kids should be vaccinated for tetanus prior to castration. A veterinarian should perform surgical castration. Bucks can also be neutered using a small Burdizzo® emasculator, a tool that crushes the testicle cords. Goat keepers often castrate kids using an elastrator, a device that stretches a rubber ring so that it can be slipped over the scrotum above the testes. The rubber band cuts off circulation and the scrotum withers and drops off.

Tattooing or Microchipping

As soon as a kid is born, place a neckband on it with the name of its mother and its date of birth. Note the birth in your reproduction records along with the name of the sire. One to four weeks after birth, kids can be tattooed with a permanent ID number. Tattoos are preferable to ear tags or neck chain IDs, which goats are likely to chew or pull off. If you plan to register a kid with the ADGA, it must be tattooed. Tattoos help identify a goat long after you have sold it and may aid in the recovery of a lost or stolen animal.

Tattoo sets are available from farm supply stores and goat suppliers. Get one with a ¼- or ⁵⁄₁₆-inch die and green ink. Most goats are tattooed on the ear, except for LaManchas, which are tattooed in the tail web.

TIP: Supplies for tattooing:

Tattoo tongs with numbers and letters

Ink, paste or roll-on, preferably green

Rubbing alcohol

Toothbrush

Stand behind your goat facing the goat's rear end. The ear on your right side is the goat's right ear; the ear on your left side is the left ear. The right ear is tattooed with a unique tattoo sequence, a series of letters and numbers identifying your farm. This sequence is assigned by the ADGA when you become a member. The tattoo in the left ear identifies the specific goat. The year of the goat's birth is indicated by a letter of the alphabet designated by the ADGA: 2007 – X, 2008 – Y, 2009 – Z, 2010 – A, 2011 – B. This is followed by a herd identification number showing the order in which the kid was born. For example, the 22nd kid born in 2010 would be tattooed with "A22."

Fit the correct letters and numbers into the tattoo tongs and test them by punching a piece of paper. Clean an area on the inside of the ear with rubbing alcohol — avoid veins or freckles. Rub the ink over this area. Placing the smooth rubber side of the tongs against the outside of the ear, puncture the ear firmly. Tattoos should be placed so they read right way up. Using the toothbrush, rub more ink over the punctured numbers with the toothbrush for about 20 seconds. Tattooing is done quickly and is usually painless.

To read tattoos on brown or black-skinned ears, take the goat into the shadows and shine a flashlight through the back of the ear. The green ink will show up clearly.

Microchips inserted under the skin are becoming a popular way to identify goats because they are less vulnerable to tampering. Some goat breed associations require microchipping. The microchip is usually inserted in the head near the ear or in the tail. A microchip must be read with a scanner and can sometimes migrate under the skin.

Wattles

Wattles are one or two long, thin bags of skin that hang down from the neck and can appear in any breed. No useful function has ever been identified for them. They are typically removed to give the dairy goat a cleaner neckline. Wattles can cause problems when they get caught on fences and torn or are sucked and chewed on by other goats. Wattles can be snipped off young kids with a pair of sharp, clean scissors. The wound heals without a scar.

Health Problems of Kids

Kids that are fed colostrum, kept in a clean dry environment out of drafts, given outdoor exercise, and fed carefully will usually grow up with few health problems. The most common health problems are diarrhea and constipation. Ordinary diarrhea can be caused by eating too fast, consuming too much milk or milk that is too cold, or changing feeds too quickly. Reduce feed and offer it at more frequent intervals. If the diarrhea continues, try replacing half of the kid's milk with water for 48 hours. In cases of severe diarrhea, or scours, withdraw all milk and feed for 24 hours and give the kid warm chamomile or black tea with charcoal or oak bark powder mixed in it. Kaopectate can also be administered to stop the diarrhea. When the diarrhea stops, resume the kid's regular feeding program gradually. Continued severe or bloody diarrhea may indicate a bacterial infection requiring veterinary attention.

A kid with diarrhea can quickly become dehydrated. If the kid's skin sticks together when pinched, give it an electrolyte immediately to replace lost body fluids.

TIP: Electrolyte mix

1 quart purified water

¼ teaspoon baking soda

½ teaspoon salt

2 tablespoons honey or white Karo® syrup

Mix well. Make up a new batch daily and feed in place of milk, doubling the usual quantity. Return to regular milk after one and a half to two days.

Signs of constipation are difficulty passing feces, straining, and hard, dry pellets. It is usually caused by overfeeding, ration that is too coarse, or lack of water. Exercise and plenty of water help, as does a laxative such as bran. In extreme cases, an enema of warm, soapy water can be given.

Kids overly infested with worms will become anemic, exhibiting pale mucous membranes and poorly conditioned coats, lack of appetite, and diminished growth. Kids are generally wormed for the first time at 6 to 8 weeks old. For younger kids, consult your veterinarian.

Weakness and a staggering gait, together with diarrhea, are signs of enterotoxemia, caused when Clostridium bacteria in the digestive system multiply rapidly and release toxins. Enterotoxemia occurs in particularly well-nourished kids.

Crooked limbs and difficulty walking are signs of rickets, a vitamin D deficiency that occurs when kids do not get enough exposure to sunlight.

Floppy kid syndrome (FKS) affects kids between 3 and 10 days old. Symptoms are muscular weakness and depression, progressing to flaccid paralysis and often death. The abdomen is distended and may "slosh" if the kid is gently shaken. The cause is still unknown, but it is thought the overconsumption of rich milk triggers the development of certain microorganisms in the digestive tract, resulting in acidosis. If detected early, it can be treated by oral administration of sodium bicarbonate and/or the tube feeding of electrolytes. More severe cases can be treated with isotonic intravenous 1.3% sodium bicarbonate solution.

Chapter 10

Goat Health

Goats have a reputation for hardiness and endurance, but they are susceptible to illness if they are not fed or cared for properly, or if they are subjected to undue stress either because of the weather or harassment by other goats, predators, or humans. Stress particularly weakens the immune system of a goat, and a traumatic experience can result in pneumonia.

You can do a lot toward keeping your goats healthy by:

- Providing the right amounts and proportions of food and a steady supply of drinking water

- Keeping their surroundings clean — clean stable, clean feed, and clean water. Many goat diseases are transmitted when goats come into contact with parasite eggs, bacteria, or viruses that have fallen on the ground from other goats

- Providing an airy but draft-free, dry stable

- Making sure goats have enough exercise

- Parasite prevention through regular testing and worming

- Regular skin and hoof care

- Consistent washing of udders, dipping of teats, and observation of hygiene during milking

- Proper vaccination program

- Preventing injury

- Daily observation of your goats and quick response when symptoms appear

Consult your veterinarian or agricultural extension office to learn what particular diseases and parasites, if any, are problems in your area. They may recommend parasite testing or a vaccination program to prevent the spread of diseases that have affected local goat populations. The soils in your area may be deficient in one or more essential minerals, which make supplements crucial to the health of your goats.

Goats' curiosity and their habit of putting their forelegs up on trees and fences to eat make them susceptible to injury from damaged fences, wires, and sharp edges. *Chapter 3 discussed how to construct enclosures and shelters so goats do not hurt themselves.* Limit the grazing areas of dairy goats so their soft, low-hanging udders are not scratched and cut by underbrush. Bucks and meat goats do not have this problem. Goats can also be injured when vying for dominance in the herd or when horned goats are kept with hornless goats. Keep an eye out for excessive aggressive behavior or horseplay and isolate either the victim or the troublemaker until she can be reintegrated with the herd. *Chapter 7 discussed goat behavior.*

Signs of health and sickness in goats

Healthy Goat	Sick Goat
Eats well	Does not eat
Chews its cud	Does not chew its cud, and rumen is inactive
Lively, alert, and sociable	Goes off by itself and lies down a lot
Smooth, shining coat (goats kept outdoors or in the cold tend to have shaggy coats)	Rough coat, bristling coat, hair falls off, bare patches
Walks and jumps with ease	Limping, not putting weight on one leg, swollen joints, stiffness
Well nourished but not fat	Swollen body, bloated; skinny and malnourished, arched back
Clear, clean eyes	Runny eyes
Cool dry nose	Nose is hot, runny discharge from nose or muzzle
Mucous membranes are soft and moist but not runny	Pale mucous membranes
Droppings are firm little balls	Diarrhea (intermittent diarrhea can be detected by little pieces of dried feces in the tail and the area just under it)
Urine is brown and clear	Blood in urine
Body temperature is 102.2 degrees F to 104 degrees F (39 C to 40 C)	Elevated temperature
Pulse rate is 40 to 80 heartbeats per minutes (kids' are slightly faster)	Pulse is racing, or slow
BUCKS	
Housing	30 sq. ft.
Stall partition	4 ft.
Yard	70 sq. ft.
Fence height	5 ft.

Isolate New Goats From Your Herd for 30 to 60 Days

Goats can easily pick up infectious diseases and parasites at live-stock auctions and fairs. Even when you buy a goat from a reputable farm, it might carry a disease. Infectious diseases have an incubation period, and symptoms may not appear right away. Once an infection or parasite spreads to your herd, it is hard to bring it under control. To keep your herd healthy and safe, quarantine new goats, and do not allow them to interact with your other goats through fences for 30 to 60 days.

During the quarantine period, vaccinate the new goats for tetanus and enterotoxemia unless the seller has provided you with reliable vaccination records. Treat them for external parasites; worm them; watch for signs of respiratory illness and pinkeye. Observe them carefully, and if you have any doubts about their health, call your veterinarian to come and examine them.

Goats should also be quarantined after they have been boarded with a veterinarian, visited another farm for stud service, or taken to a fair or show ground.

Treating Sick Goats

You can perform some medical treatments yourself, such as first aid for cuts and scratches, the administration of worm medicine, some vaccinations, and even the lancing of abscesses. For most sick goats, however, you will need a veterinarian to diagnose the disease and prescribe the correct medication. The veterinarian can show you how to continue routine treatment on your own.

Not all veterinarians are experienced in treating goats. Ask for recommendations from other local goat owners. Get to know your veterinarian before you have a goat emergency or illness, perhaps by asking him or her to examine your herd and stable and make recommendations about local health problems. Post the veterinarian's telephone number where it is easily visible. If possible, get the number of a second veterinarian to call in emergencies when the first one is unavailable. Establish relationships with experienced goat keepers in your area who can act as mentors and whom you can call for advice when you have problems with your goats.

As soon as you detect symptoms of disease, isolate the sick goat in a clean, quiet stall away from the herd, and provide her with her own supply of feed and water. Because separation from the herd can cause stress in a goat, locate a companion goat somewhere nearby where she can be seen and heard but will not be exposed to the sick goat. Goats appear to become discouraged quickly when they are seriously ill. Give your sick goats lots of attention, stroke it, talk to it, and hand-feed it treats. This kind of sympathetic support is an important factor in its recovery.

When administering medications, follow the directions carefully. Many livestock medications are formulated for larger animals and may not list the correct dosage for goats on their labels. Only a handful of animal medications have been officially approved for use in goats. Your veterinarian will know the right dosage and can instruct you on the best way to give the medicine to your goat.

Vaccinating Your Goats

A whole range of vaccines are available for goats, but some are expensive and not all of them are necessary. Follow the recommendations of your veterinarian, local agricultural extension office, and other local goat owners regarding vaccinations against diseases that are threats in your area.

Goats everywhere should be vaccinated against tetanus and enterotoxemia (overeating disease) once a year. These vaccinations are available as a combination in a single vaccine. All your goats should receive an initial vaccination followed by a booster three to four weeks later and an annual vaccination after that. Pregnant goats should receive their annual vaccination one month before they give birth. Kids should be given their initial vaccination when they are one to two months old, followed by the booster three to four weeks later. Goats that are fed large amounts of concentrated feed may need to be vaccinated every six months because they are more susceptible to enterotoxemia. *This disease is discussed in more detail later in this chapter.*

TIP: How to give an injection to a goat

You can quickly learn to administer routine vaccinations yourself. Use a clean, new syringe for each session and a clean needle for each goat. Use sharp 16- and 18-gauge needles in ½-, ⅝-, or ¾-inch lengths. Shorter needles are used for subcutaneous injections.

Swirl (do not shake) the vaccine bottle to mix the contents. Pull back the plunger of the syringe a little further than the dose you plan to inject. Holding the vaccine bottle upside down, poke the needle through the rubber. Depress the plunger to inject air into the bottle. Pull the plunger back a little farther than the required dose, and then gently squeeze the excess back into the bottle to get rid of air bubbles.

Always use a new needle to draw vaccine from a bottle to avoid contaminating its contents. If you want to reuse needles to vaccinate your goats, leave a clean needle stuck in the bottle and detach the filled syringe. Reattach the syringe to refill it, then remove it, and replace the used needle to give the injection.

Subcutaneous (under the skin) injections: Most vaccines are given subcutaneously in the neck, over the ribs or in the hairless area of the armpit. Choose an area of clean, dry skin, and swab it with alcohol. Pinch the skin and lift up a little "tent" then slide the needle in and slowly depress the plunger. Withdraw the needle and rub the area to distribute the vaccine.

Intramuscular injections: Some antibiotics are given as intramuscular injections, usually into the muscle at the side of the neck. Have someone restrain the goat, and quickly and smoothly plunge the needle deep into the muscle. Pull back on the plunger about ¼ inch. If you see blood, you have penetrated a vein and must withdraw the needle and try again. Depress the plunger slowly and withdraw the needle.

First Aid for Wounds

Wounds should be disinfected with tincture of iodine or a coating of pine tar. To stop excessive bleeding, apply pressure with a clean, folded towel. If a wound is inflamed, wash it or bathe it with chamomile tea. Scratches on the udder can be treated with antibiotic ointment. For abscesses, apply a dressing or icthyol ointment. A veterinarian should stitch deep or gaping wounds. Dust with a fly-repellent powder or spread a fly-repellent ointment around the wound to keep flies off.

Common Diseases of Dairy Goats

Many diseases of dairy goats are caused by nutritional deficiencies or occur because the goat's natural resistance is weakened in some way. Proper care and feeding is the best defense.

Some diseases have similar symptoms. Unless you are certain you know what is wrong with your goat, do not waste time trying to be a goat doctor. Consult a veterinarian or a seasoned goat keeper when one of your goats is behaving abnormally or showing symptoms of illness. Early treatment is most effective and prevents the spread of disease to the rest of your herd. Do not wait to call your veterinarian as a last resort, when you have tried everything, and your goat is at death's door.

The following descriptions of common goat diseases will acquaint you with some of the signs and symptoms to watch for, as well as some of the mistakes to avoid.

Caseous Lymphadenitis (Abscesses)

Abscesses are a common chronic disease of adult goats. Bacteria enter through small cuts in the skin and localize in the lymph nodes. Abscesses arise from the lymph nodes, particularly around the head, neck and shoulders, and front of the rear knee joints. They can eventually cause the affected animal to become emaciated and die if internal abscesses interfere with vital organs. If the abscesses rupture, the greenish pus quickly transmits the bacteria to other goats in the herd.

Goats should be isolated as soon as the condition is detected. The abscesses should be lanced after they start to come to a head near the surface of the skin and the pus carefully collected and disposed of. Follow up with four daily shots of penicillin and flush the wound with an antiseptic solution until healed. The animal should be isolated during treatment, and the area around the wound washed and dried before returning her to the herd.

Eradicating caseous lymphadenitis from a herd can only be done through a planned program of raising offspring in separate facilities and then disposing of the infected animals. The use of an autogenous bacterin prepared by a laboratory is thought to be helpful in reducing the incidence of disease. No commercial vaccine is currently available.

Contagious Ecthyma (Sore Mouth)

Sore mouth is caused by a resistant virus that produces scabs about the lips and gums. The virus is transmitted in the scabby material, which may remain viable in the soil for a long period. This disease is more serious in kids because the soreness prevents them from eating normally and because the infection may spread to their mothers' teats when they nurse. Immunity develops after the initial infection. A vaccination program can prevent the disease.

The purpose of treatment is to prevent a secondary bacterial infection. Gently remove the scabby material using gauze soaked in hydrogen peroxide, and then cover the area with zinc oxide or similar ointment. Wear plastic gloves because it is transmissible to humans.

Mastitis

Mastitis is an inflammation of the udder. It may be acute or chronic. Most cases are caused by streptococcus or staphylococcus organisms. The udder may appear hot, painful, tense, and hard. A wide spectrum antibiotic may be needed or simple penicillin may be effective. The disease can be cured if treated early.

Udder edema

Udder edema and congestion is commonly observed in high producing dairy goats during the late dry period and after parturition (giving birth). Although the problem cannot be totally controlled, limiting the use of sodium (salt) and potassium (good sources are alfalfa hay and cane molasses) as well as high-energy feedstuffs, such as corn meal, in the dry period is helpful. Corn meal should be limited to about 20 percent of the ration. The total ration dry matter should contain about 0.2- to 0.3-percent sodium and 0.7-percent potassium. Although a lower energy and higher fiber ration is needed for the dry does, lactating does need higher energy feedstuffs in their rations with adequate amounts of good quality forages.

Polioencephalomalacia (PEM)

PEM is a thiamine deficiency most commonly occurring in kids 2 to 6 months old who are fed concentrate, or feed that contains a high proportion of nutrients and is low in crude fiber content (less than 18 percent of dry matter). Feed high in grain and low in forage can suppress the normal production of thiamine by the digestive system. Feeding moldy hay, feed high in molasses, or sudden changes in feed can trigger PEM. The goat exhibits blindness, depression, lack of coordination, pressing the head against walls and fences, and convulsions. In severe cases, the goat falls into a fatal coma. Thiamine injections should be given as soon as the disease is detected, and the diet should be adjusted to include more forage and less grain.

Pinkeye

Pinkeye is a rapidly progressing eye infection caused by bacteria. Tall grasses, flies, and dry, dusty conditions seem to contribute to its spread. The goat begins closing its eye. This is followed by increased tearing, inflammation, and cloudiness and ulceration of the cornea that can cause temporary blindness.

Infected goats should be separated from the herd and treated with topical ointment, or in some cases, with antibiotic injections administered by a veterinarian. Quarantining new goats can prevent the spread of pinkeye because it does not always appear right away.

Swollen joints

Various rheumatic illnesses affect goats in damp, cold environments, and they are not well understood. One type, caprine arthritic encephalitis (CAE), is caused by a virus that is apparently passed from mother to kid through colostrum. In the early 1990s, researchers at the Washington State University found that 80 percent of the dairy goats they tested carried antibodies for CAE. An effort has since been made to eradicate it.

The disease progresses slowly and eventually manifests as severe swelling in the joints and apparent inflammation of the udder. There is no cure for CAE. Application of salves and heat to increase circulation in the affected areas can ease pain. Some goat keepers prevent carriers of CAE from further breeding. Kids can also be separated from the mother at birth and fed with colostrum from a healthy goat or cow. Colostrum can be milked from the infected mother and cleared of the CAE virus by heating it to 131 degrees F (55 degrees C) for 60 minutes.

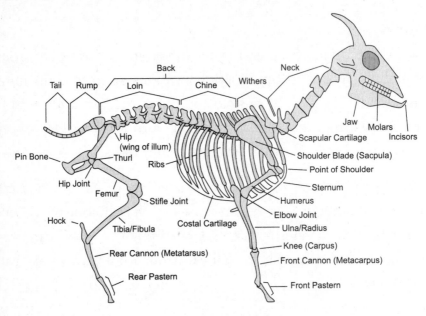

Example of a goat skeleton

Digestive Diseases

Diarrhea (scours)

Diarrhea is often a response to mistakes in feeding, such as an over-high albumin content, an abrupt change in feed, too much concentrated feed, or feed that is rotten, fermented, or frozen. It can also be caused by infectious diseases and by internal parasites. Feed-related diarrhea is typically brothy and liquid, with a powerful smell. Watery stool mixed with blood indicates an infection of some kind. Intermittent diarrhea is a sign of worms.

The treatment for feed-related diarrhea is to withdraw concentrated feed and give plenty of water and branches of fir, spruce, and pine. If the goat has a fever or other symptoms, or if you suspect infection, consult a veterinarian. If diarrhea persists, check fecal samples for worms. Prolonged diarrhea leads to dehydration and a rough coat. Kids should be treated immediately because diarrhea can be fatal to them.

Enterotoxemia (overeating disease)

Enterotoxemia is caused when a sudden change in feed or overeating by hungry animals stimulates the bacteria and causes them to grow rapidly and release a toxin in the intestines. Regular feeding and vaccinating with the vaccine Clostridium perfringens type C and type D toxoid will prevent this disease. In adult animals, the symptoms are depression, intoxication, and poor coordination; in kids, it often just manifests as sudden death.

Bloating

Bloating occurs when a foreign object blocks the esophagus and the goat cannot burp up the digestive gases in its rumen, or when rapid eating or overeating of concentrated feed or damp green clover produces fermented foam in the rumen. Bloating in a kid is caused by faulty milk digestion. The left side of the goat bulges out, and the goat is obviously in pain. The expanding rumen restricts lung function and can cause respiratory failure. Bloat is potentially fatal.

To treat for bloating, remove the blockage from the esophagus by locating it from the outside and massaging it downwards. If the bloat appears to be the result of fermenting foam, instill a foam-destroying preparation, such as vegetable oil or silicone, from the vet. You can also stimulate the goat to produce and swallow more saliva by running a string gently back and forth through its mouth. Saliva breaks down the foam. As a last resort, you can release pressure by puncturing a hole in the rumen with a tool called a trochar or with a penknife. However, this will bring only minor relief for bloat caused by fermenting foam in the rumen). This should only be done in emergencies when you cannot reach a veterinarian, and the goat appears to be in crisis, as it can result in an infection.

Acid rumen

Acid rumen occurs if a goat rapidly ingests a large quantity of sugar-rich food such as bread, sugar beets, or concentrated food. The goat seems apathetic, hangs its head, and looks as though it is drunk. Bloating can also occur. To treat acid rumen, supply the goat with plenty of water. If the goat cannot stand up, call the veterinarian immediately.

Store concentrated food where it is secure from goats, and feed sugar-rich foods to them in gradually increasing quantities so they can become accustomed to it.

Foot scald

Foot scald is a moist, raw infection of the tissue between the toes. It is caused when *Fusobacterium necrophorum*, a bacterium that thrives in soil and manure, invades a scratch or lesion in a hoof that is constantly immersed in mud or warm muck. Foot scald typically occurs in only one of the front feet.

Foot rot

The germ that causes foot rot, *Bacteroides nodusus*, also thrives in wet, muddy areas where air is poorly circulated, and it multiplies in the presence of *F. necrophorum*. *B. nodusus* penetrates the deeper layers of the skin and releases an enzyme that causes surrounding tissue to liquefy.

A goat with foot rot limps noticeably. Symptoms include a grayish, cheesy discharge and foul odor with lameness and intense pain. Treat by carefully trimming away the rotten area and treating the infected area with 10 to 30 percent copper sulfate solution, a suitable ointment, or other treatment as prescribed by your veterinarian.

Once your herd is infected with foot rot, treating and eradicating it is a lengthy process. You can prevent foot rot by housing your goats in a relatively dry area and by maintaining their properly trimmed hooves. Trim the hooves of new goats and isolate them for three weeks before integrating them in your herd. Similarly, isolate any goat that leaves your farm and returns, including goats that have been to shows or fairgrounds or boarded at a veterinary facility.

Problems of Pregnancy

Ketosis

After a kid is born, the doe's milk production sometimes increases so rapidly that it cannot eat enough feed to keep up, which causes the body to break down its own body fat. The breakdown of fat releases ketones that can be toxic in large quantities. The goat's feeding slows down, its milk production falls off, and its breath and urine have a sweetish odor. The goat is apathetic and has a rough coat. In severe cases, the goat can become comatose and die.

A veterinarian can treat ketosis. To prevent it, begin gradually increasing a pregnant goat's portion of concentrated food about two weeks before her kid is born. Start with small handfuls and gradually increase it to 1.1 pounds per day by the time the kid is born. Feed the goat high-quality feed so she gets the maximum amount of nourishment.

Toxemia during pregnancy

Toxemia can arise near the end of a pregnancy, particularly when a goat is carrying twins or triplets. The cause is similar to ketosis. The growing babies require more and more energy while the expanding uterus occupies more space, pushing on the rumen and forcing the goat to eat less and less. As the goat's body breaks down fat reserves, toxic quantities of ketones are released, and the goat may appear blind, stop eating, and fall into a fatal coma. In this case, there is little a veterinarian can do to save the goat.

Toxemia is more likely to occur in a fat goat, so it is best to keep milk goats slightly thin and not feed them concentrated food during their dry periods before they are bred. Goats are less prone to toxemia when they get ample exercise. Feed them the best possible roughage during the last few weeks of their pregnancies, but do not start increasing their ration of concentrated food until the last two weeks of pregnancy.

Milk fever

Milk fever is caused by a calcium deficiency brought on by milk production right after the birth of kids. The goat's body fails to draw on calcium reserves in its bones, especially if the goat has been fed calcium-rich alfalfa or concentrate during the final weeks of pregnancy. The goat exhibits a weak, limping gait before the birth and may sink into a coma and die after the birth. A veterinarian can rescue a goat even in acute circumstances by giving a calcium infusion. Avoid overfeeding calcium in late pregnancy, and compensate for calcium-rich food with increased amounts of phosphate in the diet.

Johne's Disease

Johne's (Yoh-neez) disease is an incurable, contagious, antibiotic-resistant bacterial disease that infects the intestinal tracts of ruminants. Afflicted goats are dull, depressed, and thin, and they eventually die.

Poisoning

Goats can be poisoned by eating or licking lead-based paints, ingesting pesticides and fertilizers, or eating poisonous plants inadvertently mixed in with chopped feed. They may eat toxic ornamental plants along fences or wilted leaves of trees in which toxins are concentrated during the fall. Symptoms of poisoning are vomiting, frothing, staggering, rapid or labored breathing, changes in pulse rate, labored breathing, cries for help, and sudden death. If you realize early enough that a goat has been poisoned, you may be able to induce vomiting with a drink of warm salt water or with 2 tablespoons of salt placed on the back of the tongue. A veterinarian may be able to administer an antidote to the poison if it can be identified.

Respiratory Illness

Poor air quality, especially with a high content of ammonia fumes from urine, and cold, damp living quarters can bring on inflammations of the respiratory system. Goats lose their appetite, run a fever, breathe hard, have a nasal discharge, sneeze, or cough.

It is easier to prevent respiratory illness than to treat it. Maintain a well-ventilated, but draft-free, stable and dry, warm sleeping quarters. In winter, a cold, dry stable is preferable to a warm, damp one. Avoid stressing your goats during cold weather by changing their stables or feed unnecessarily.

Skin Problems

Goats are plagued by fungus infections and lice and mites, particularly if they are kept in a stable where they are not able to groom themselves, scratch against fence posts, rub, and lick themselves. The goat constantly tries to scratch itself, hair falls out, and scales, scabs, and bald patches appear.

Goats have healthier skin if they spend time outdoors. If kept in a stable, they should be brushed regularly with a stiff brush.

Lice

Goats that are twitchy and fidgety and have dull coats are probably infested with lice. Small amounts of lice do not cause problems for healthy goats, but a goat with a serious infestation will rub itself against every available object, lose hair, and have dandruff and dry skin. Lice can be controlled by dusting, spraying, or dipping. Your veterinarian can recommend a louse powder suitable for dairy animals. The entire herd must be treated at once. An old remedy is two parts lard to one part kerosene rubbed on the goat. Fresh air, rain, and sunlight prevent lice from becoming a serious problem.

Mange

Mange is caused by tiny mites that burrow into the skin. Symptoms of mange are irritated, flaky skin followed by hairlessness and thick, hard patches of skin. Demodectic and psoroptic ear mange are specific to goats, while sarcoptic mange affects all species of animals. Mange can be treated with medications available from a veterinarian.

Screwworms

Screwworms are present in the South and Southwest United States despite eradication programs. Screwworm flies lay eggs on open wounds, and the eggs hatch into hundreds of larvae that feed on the living flesh for five to seven days. If a goat has a foul-smelling wound with larvae in it, contact your veterinarian immediately.

Ringworm

Fungal skin infections such as ringworm require treatment. Treatment includes using a solution of glycerin or tincture of iodine. Apply daily treatment of a mixture of equal parts tincture of iodine and glycerin or a 20-percent solution of sodium caprylate to the lesion until it disappears. The antifungal activity of thiabendazole may provide a useful treatment.

Internal Parasites

Nearly all goats, even those that are healthy, harbor various kinds of worms that inhabit the stomach, intestine, liver, or lungs. Eggs are excreted and develop into infectious larvae either on the ground or in an intermediate host. These larvae are ingested by feeding goats and grow into worms inside the goats' bodies. Goats are susceptible to illness caused by these parasites whenever their resistance is diminished by stress, heat, poor nutrition, or another health problem. Goats also harbor coccidia, single-celled parasites that do not harm adult goats but can cause fatal intestinal illness in kids.

Symptoms of internal parasites are weight loss, rough coat, paleness of the mucous membranes of the eye and mouth, repeated diarrhea, and decline in milk production. Goats suffering from internal parasites often arch their backs, and there may be swelling under the jaw. Lungworms can cause coughing and a nasal discharge. The condition can be diagnosed by examining the feces of the goat under a microscope to determine what type of eggs are being excreted.

The best way to prevent illness is to maintain the goats' resistance by providing a good diet with a high content of minerals and vitamins. Good hygiene and pasture management help to control the numbers of parasite larvae ingested by the goats. Implement the following recommendations to decrease the chance internal parasites will harm your goats:

- Keep water and feed from being dirtied by goat droppings.

- Clean the feeding shelves and water containers often.

- Try to provide fodder, such as hay and leafy branches, free of parasite larvae.

- Replace litter straw often and construct floors and yards so they are easy to clean.

- Collect green feed from fields that have not been used as goat pastures or been fertilized with goat manure in the past year.

- When rotating goats from one area of pasture to another, leave the grazed areas to rest for three to six months. Large numbers of larvae may survive even longer in humid climates. Ideally, the goat pasture should be used as a hay field in between grazings, or used to pasture cattle or horses because they do not carry the same parasites as goats. If possible, there should be a two-year interval before an area is used again to pasture goats.

- Drain or avoid swampy pastures. The large liver fluke thrives in host snails that live only in swampy areas.

- Limit the number of goats in a pasture to six or seven per acre.

Worming your goats

Commercial worming preparations such as moxidectin, ivermectin, levamasole, fenbendazole, and albendazole have proven effective in controlling internal parasites. Before worming, do a fecal analysis to identify which parasites are present and in what numbers. Your veterinarian can suggest the most effective wormer and recommend the right dosages.

Worming preparations (anthelmintics) come as boluses (large pills), liquids, pastes, gels, powders, and crumbles. Boluses are popular, but many goats balk at swallowing them. They can be hidden in a wad of peanut butter or administered with a balling gun. Some goat keepers refuse to use boluses because they can choke a goat.

If using a paste, make sure the goat's mouth is completely empty. Put the paste in the back corner of the mouth on the left side. Gently hold the goat's muzzle closed and massage her throat until the paste has been swallowed. Drenching is administering a liquid using a bottle. It can be tricky but is a necessary skill for a goat keeper. Squirt the liquid in the left-hand corner of the mouth, stopping at regular intervals to allow the goat to swallow. To avoid getting liquid in the lungs, keep the muzzle level and never raise the head.

Goats in rotating pastures should be wormed just before they are moved to a new pasture so they drop their eggs before the move. The most effective times to worm your goats are in the last month of pregnancy, a day after kids are born, at the beginning of the spring/summer, and just before the goats are returned to the stable for the winter. Goats that remain in the same pasture year-round should be regularly checked and wormed. It is difficult to control parasite larvae without resting pasture. If possible, goats should be confined in the stable for two days after they are wormed so they drop the eggs there. Goats kept in a stable all the time do not need worming if their feed and water is kept clean.

Humans should not consume milk from newly dewormed goats. Read the label, ask your vet, or look online to see how long you must wait before using the milk again.

TIP: Diseases that can spread from goats to humans

- **Toxoplasmosis:** A disease of cats, caused by a protozoa that can result in miscarriage in goats and humans. The symptoms are so mild in adult goats that they are often not noticed, but the microbes can be excreted in milk and can injure fetuses and infants who drink raw milk. Pasteurizing or boiling goat's milk before human consumption prevents spread.

- **Tuberculosis:** Tuberculosis arises in various organs of goats and can be transmitted in milk. Tuberculosis has been all but eliminated in the domestic animals of most countries.

- **Brucellosis:** Brucellosis is still widespread in the Mediterranean and in tropical and subtropical countries. It causes undulant fever in humans and miscarriage in goats. Goats can carry the disease without exhibiting symptoms. Pasteurizing goat's milk prevents transmission.

- **Chlamydia:** A common cause of miscarriage in goats in the United States, chlamydia can spread to humans and cause miscarriages. Pregnant women should avoid contact with goats during kidding time and should wear masks and gloves when handling goats.

- **Q Fever:** A cause of miscarriage in goats and sheep, the disease can spread from cows to goats. Humans can contract it by inhaling contaminated dust, consuming unpasteurized milk, and coming into contact with miscarriage material. Human beings experience flu-like symptoms.

- **Rabies:** Pastured goats can get rabies if a rabid animal ,such as a rat, raccoon, or vampire bat, bites them. Goats with rabies exhibit abnormal behavior but are not necessarily aggressive. Transmission to humans can only occur if the rabid goat's saliva enters an open wound. Contact your veterinarian if you suspect that a goat has rabies or has been bitten by a rabid animal. Like humans, goats can undergo a series of injections to prevent rabies.

Chapter 11

Running a Dairy Goat Business

eeping dairy goats as a commercial enterprise is not the same as keeping a few dairy goats to provide milk for a family. On a small farm where pasturage and shelter are readily available, dairy goats provide an opportunity for sustainable diversity and occasional income. On a vegetable farm, goats provide fertilizer and compost, and they eat discarded leaves and residue while supplying milk for the family or for raising kids, calves, piglets, or lambs. Goats can be pastured along with cattle or sheep, and their browsing keeps pastures clear of brush and young trees. Raw goat's milk can be sold to neighbors to feed pets and livestock, and surplus kids can be sold for meat or dairy stock. If you keep a buck, you can earn extra income through stud fees. You can even rent out your bucks to neighboring farmers to clear their pastures. Does have to be kept where they can be milked regularly and where they will not injure their udders on rough undergrowth.

To succeed as a commercial enterprise, your dairy business must bring in enough income to cover everyday expenses and gradually pay off your initial investment, as well as making a profit. A commercial dairy goat farmer faces many challenges, including finding a steady market for goat's milk, cheese or goat's milk products and staggering breeding so the does supply milk year-round. The milk and cheese must be safely stored and transported to their final destinations.

As you have seen from the earlier chapters in this book, successful goat keeping involves a range of skills, including carpentry and fence building; selecting and storing good-quality hay; feed mixing; grooming; and milking. From time to time, you will find yourself in the role of animal psychologist, matchmaker, midwife, stable hand, and amateur veterinarian. The more you can do yourself, the less you will have to pay for outside assistance.

Before you make the decision to invest in dairy farming as a commercial enterprise, make a thorough and complete business plan. Your plan should consider labor, the marketing outlook, processing options for your milk, regulations, budgeting, and economics.

Labor

If you intend to keep more than a handful of does, you will need someone to help with the work. Goats must be milked and fed twice a day, seven days a week. Are family members willing to support you? Will you be able to hire enough competent and reliable workers to keep your dairy operation running smoothly?

Agricultural experts vary in their estimates of the amount of labor required to maintain a dairy goat herd. In a 1989 publication, *Economics of the Dairy Goat Business*, Dr. Robert Appleman of the

Minnesota Extension Service gave the following estimate for labor required for dairy goat operations:

- Milking: 25 does/person/hr. (305 days)
- Setup and cleanup: 40 min. daily
- Manure handling and bedding: 25 min. daily
- Feeding hay and grain: 30 min. daily
- Heat detection: 30 min./day for six months
- Breeding: 20 min. each for two breedings
- Miscellaneous care: .5 min. daily per doe

Total labor per doe in Appleman's budget is 34.7 hours per year, 70 percent of which is spent milking. A labor budget developed by Pennsylvania State University estimated labor to run a 100-doe facility as 22 hours per doe per year, and a Rutgers Cooperative Extension budget estimated 13.6 hours per doe per year for a 100-doe herd. To get a realistic idea of the amount of labor involved, visit a dairy facility similar in size to the one you plan to operate and interview the owners. Observe how feeding, daily milking, and other activities are organized and managed.

You can reduce labor by organizing and building your goat facility so work can be done efficiently and cleanup is easy, but you cannot reduce the amount of time needed to milk each goat. Once you have determined how much assistance you will need, investigate the local labor market. Ask other farmers and your agricultural extension officer about hiring practices and wages. If possible, acquire copies of hiring contracts local farmers use. If you are taking out a loan to fund your business, your lender may require you to purchase insurance to cover liability for injury to an employee. Your insurance company may require an additional premium to cover employees.

Marketing

How do you plan to make money from your dairy goat business? What kind of products do you intend to sell? What kind of demand is there for those products? What prices can you sell them for? Will that price cover your production costs and still give you an income?

If you plan to sell goat's milk, you need to find a reliable buyer. In many areas of the United States, there are no companies that process goat's milk for commercial sale. Even where a processor does exist, it may not need additional milk producers. If you are unable to sell your goat's milk to a processor, it may be feasible to sell it to individuals raising baby animals or as pet milk at local farmers markets. Most states have prohibitive restrictions on the sale of milk or milk products directly to individuals for human consumption. Contact the agency responsible for dairy regulations in your state. The American Dairy Goat Association (ADGA) website lists the contact information for state agencies (**www.adga.org**).

Any dairy selling raw milk to be processed into Grade A milk or milk products must meet a number of requirements to receive a Grade A dairy permit. The permitting process is carried out by state regulatory agencies that send out inspectors and regularly test the milk for bacteria and drug residues. State agents typically work with a new dairy farmer during the design and building of the dairy facility to ensure requirements are fulfilled. Requirements vary from state to state but typically include the building of a separate room for milking and milk storage. The additional cost of building these facilities, application fees, and the time it takes to complete the application and inspection process must be factored into your business plan.

Individuals who want to purchase raw milk for their families sometimes circumvent legal restrictions by purchasing part ownership in a goat. The goat stays on the farm, but the family regularly receives its share of the milk and pays for part of the goat's upkeep. Information on goat share contracts can be found on the website for the Campaign for Real Milk (**www.realmilk.com**). Raw milk can also be sold if it is to be fed to pets or livestock. Locating individuals who want to purchase your raw milk takes a lot more effort and time than signing a contract with a milk processor who sends a truck to pick up your milk twice a week.

You may be able to develop a regular market for your raw milk by selling it to farms that raise calves or pigs or by using it to raise your own livestock for sale as meat. It will take time to establish yourself in the local farming community and find enough customers to be able to sell your milk consistently.

Processing Your Own Milk

Rather than selling your milk, you can process it yourself and sell your own pasteurized milk, cheese, yogurt, fudge, or nonedible products such as soaps or lotions. Raw milk can be used to make cheese if the cheese is aged 60 days or more. Pasteurized milk must be used to make fresh cheeses.

To sell edible products, you will need a Grade A dairy, a commercial kitchen, and other permits or licenses. Contact your state agricultural or consumer affairs agency to learn what is required. Producing your own products involves additional labor and equipment to manufacture, package, market, and ship them.

Study the products you want to produce and make a detailed list of all the expenses involved. You will have to spend some time experimenting, practicing, and perfecting recipes and processes before you have a marketable product. Attend cheese-making classes or learn from another producer. Once you are making your products, you may be able to sell them at local grocery stores, restaurants, specialty stores, farmers markets, or online.

TIP: Requirements for a Grade A dairy

The U.S. Grade A Pasteurized Milk Ordinance (PMO), drafted by the U.S. Food and Drug Administration, states that only pasteurized milk can be sold as Grade A. Enforcement of this ordinance is under the jurisdiction of state departments of health or agriculture (Zeng and Escobar, 1995), and local requirements may vary.

Requirements for a Grade A dairy include a milking barn or parlor with a floor made of concrete or other impervious material that can be easily cleaned, and smooth, painted, or finished walls and ceilings that are sealed against dust. There must be enough ventilation to eliminate condensation, minimize odor, and provide a comfortable environment for the milker. Lighting must be adequate, and medications must be kept in a storage cabinet. Milking stands cannot be made of wood.

A separate milk room is required for cooling and storing goat milk to minimize the risk of contamination from the milking barn. The structure must be in good repair and easy to clean. The floor should slope evenly to a drain, and wash-sinks, hot water, and on-site toilets are required. Milking lines and other equipment should be of stainless steel or other smooth, nonabsorbent material. Milk storage tanks must have an efficient cooling system. Fresh, warm milk coming out of pipelines or milking buckets must be cooled to 45 F within two hours. The water supply must comply with the Clean Water Act requirements, as enforced by the EPA, and a dairy waste management system must be in place. Grade A dairies are inspected at least twice a year, and milk samples are collected periodically. (*Grade A Dairy Goat Farm Requirements*, a publication by Langston University is available at **www.luresext.edu/goats/library/fact_sheets/d04. htm.)**

There is an expanding market for soaps and other nonedible products made with goat's milk. These products do not require the same licenses and permits as edible products, but you will still need to develop your products, create a work area, and purchase additional ingredients, equipment, and packaging supplies.

Producing your own products involves additional labor to manufacture, package, market, and ship them. Each new product requires an additional investment in equipment and materials. Advertising, selling your products, processing orders, and keeping accounts is another full-time job, at least until your business becomes established. Some producers of cheese and soaps observe that the business is so demanding they have little time to spend with their animals.

Whatever products you make, they must be sold at prices high enough to pay for production costs and generate a profit. Keep an eye on your competitors' prices, but do not let them determine what you charge. Price your products to reflect their quality and the amount of work that goes into making them.

CASE STUDY: SUSAN BRITT

Owner, Soggy Bottom Farm
and Grandy's Country Store
http://grandyscountrystore.com

The rising development in South Florida left no room for Susan Britt and her husband to continue boarding horses. In 2002, they moved north to Morriston, Florida, where they bought 20 acres in horse country. Their intentions to board horses changed when Britt agreed to inherit a single goat from a friend. She found goats to be sweet animals and very easy to get along with, so she decided to raise goats.

"I started with Boer goats and then I ended up adding some Nubians," Britt said. "I do primarily raise meat goats, but I do have dairy goats as well." After significantly increasing the size of her herd, Britt soon found she had an abundance of milk. Selling the raw milk was not possible because of the laws in the state of Florida, so she decided to make goat's milk soap — right on her back porch. "I've always loved goat's milk soap," said Britt. "I started doing the research, and I found out it is not horribly difficult. But there is a technique to it.

"There are certain safety things you have to wear. You have to wear goggles. It is best to wear gloves because you do work with lye, and lye can burn."

When making the soap, you also have to use stainless steel or plastic, and work in an open area because of the caustic fumes.

After refining the soap-making process, Britt decided to sell the goat milk soap under the name Grandy's Country Store. The store is located online, but Britt wants to sell the soap through distributors as well. Britt does everything on her own: she makes the soap, dries it to let it cure for several weeks, packages it, and ships it to people who order online. "I give out a lot of samples to people and talk to them," Britt said. "I think if the soap continues to grow at the rate it has been that I will truly have to get some help."

There are two lines of soap at Grandy's Country Store: a Signature Series and a glycerin-based soap. The Signature Series is made from the milk of the goats on Soggy Bottom Farm and uses all pure ingredients. The Signature Series is made in scented and unscented varieties, with coconut oil, palm oil, olive oil, and other natural oils. "It is a beautiful soap; there is a nice texture to it. It lasts a long time, and the longer you cure it the better the bar is. I have had so much fun making the soap that now I am working on perfecting designer bars."

The packaging Britt uses on the soap is simple. A cellophane wrapper with a label on the back listing the ingredients and a small bag with a tie at the top is all it needs. Even though most of the orders are online, the marketing is solely word of mouth and relies on people trying the soap and telling other people about it.

With the knowledge and experience of someone who has been making soap for decades, Susan Britt has only been making goat's milk soap for about two years. Her interest in goats and a simpler way of life drives her to make a business out of what started as a small hobby.

For more information about Grandy's Country Store, or to purchase goat's milk soap, visit the website at **http://grandyscountrystore.com**. Ten percent of all proceeds are donated to breast cancer research.

Economic Feasibility Study

Before beginning a commercial dairy goat business, conduct a thorough economic feasibility study. Make this study as detailed as possible, and try to think of every expense. Include the cost of building or renovating facilities and installing fences and the interest you will pay on any loans you take out to start the business. You can find a number of sample dairy goat farm budgets in agricultural publications and on the Internet. *Also, see the sample budget in Appendix A.*

- Research the cost of labor, feed, hay, goats, veterinary care, and equipment in your area, and plug these costs into your budget to calculate your projected expenses. Remember that the prices of hay and feed will vary from season to season and that you will need less forage in the seasons when natural forage is available. To add a safety margin, increase your projected expenses by an additional 30 percent.

- Research the current prices for fluid milk for processing, bottled milk, milk-fed livestock, cheese, or soap. Estimate your production level. How many does are you planning to milk? How productive will they be, on average? Does in a large herd average lower production than a small herd of hobby goats.

- Create a preliminary marketing plan. How will you sell your products? Who will do the selling? How long will it take to get your products on the market? Be as realistic about production and marketing as you possibly can.

- When your budget is complete, calculate how much you will have to produce and the prices you will have to charge in order to make a profit.

Show your budget to other commercial goat farmers in the area and ask them if it seems accurate. An agricultural extension officer or a farm cooperative can advise you on whether your budget is complete. If you are taking out a loan to start your business, your lender will tell you what is required in your business plan.

Your feasibility study should include an alternative backup plan in case your intended market fizzles or does not materialize at all. For example, if a milk producer is unable to take all your milk, you might sell the rest to nearby farmers who are raising calves, or raise some milk-fed livestock of your own to sell as meat. Do not wait until a crisis happens to start developing your alternative business plan. The sooner you are able to put it into action, the faster you can cut your losses and re-establish an income stream.

A good business plan includes an exit strategy. How long do you expect to be in the dairy goat business? What will happen to this business when you retire? If your business is losing money, at what point will you re-evaluate and make changes, such as selling off some of your herd or raising prices? If you close the business, what will you do with your equipment and facilities? These eventualities may seem a long way off, but anticipating them is part of sound financial management. Knowing what to expect in the future will help you to make good business decisions today.

Your feasibility study not only helps you plan for a profitable commercial dairy goat business, but it also provides a valuable tool for measuring the progress and growth of your business. When you do your books at the end of the financial year to prepare your taxes, compare your actual expenses with the budget you created. Are you spending more or less than you anticipated? If your expenses are regularly exceeding your expectations, you will either have to increase the prices of your products or find ways to cut down your costs to maintain the profitability of your business. Sometimes you become so caught up in day-to-day management of your business that you can no longer grasp the whole picture. A review of your budget allows you to step back and see how it is performing. You may be surprised to see that you have actually exceeded the milestones you set for yourself.

Making the most of your money

A good budget is only one element of a profitable business. From the moment you begin designing your goat shelter, look for ways to maximize efficiency, economize, and minimize waste. The decisions you make now could save you a great deal of expense over the long run. Experienced goat keepers offer these recommendations:

- Plan five years ahead when designing and building your facilities. Build so you can renovate and expand without having to tear down your existing structures or make expensive modifications. Construct a larger shed now and use the extra space for storage. Instead of adding a second shelter when you increase your herd, you will be able to reorganize your existing shed to accommodate extra goats.

- Labor is expensive. Cut down on the time it takes to do chores by designing a facility that is easy to clean. Construct feeding stations and milking stands so feeding and milking can be done quickly and efficiently. Devise convenient feed storage facilities and an effective method for handling soiled bedding and manure. Visit other goat farms to see how they are organized before you start designing your own facilities.

- Secure enough capital to cover your initial expenses and get your business up and running. It could be a considerable time before your business produces enough income to turn a profit. Do not give up your day job right away. It would be a tragedy if you were to set up your business, buy your goats, and then be unable to continue your dairy goat operation because you ran out of cash.

- Put your business plan in writing and review it from time to time. Keep accurate health records and milk production records for each goat, and keep records and receipts for all your expenditures. Over time, a pattern will emerge and you will be able to anticipate future income and expenses.

- Disease and poor nutrition lower milk production and can bring your business to a sudden halt. You can avoid veterinary expenses and maximize milk production by keeping your facilities clean and organized, observing proper hygiene when milking, monitoring your goats daily for signs of mastitis, controlling parasites with good pasture management, isolating sick animals, and responding immediately to signs of poor health.

- Sell off excess kids and unproductive does immediately; it costs you money and labor every day to maintain each goat. The sale of kids is a significant source of additional income.

- Manage young does so they are ready to breed at 7 months to minimize the number of unproductive goats in your herd. This increases the lifetime production of meat and milk by your herd.

- Pay careful attention to nutrition and the cost of feed and hay. Nutritional needs of dairy goats change with pregnancy and lactation and with the seasons. Poor quality feed lowers milk production, but overfeeding is an unnecessary expense and can be harmful to the goats' health. Store hay and bedding where they will not be spoiled by moisture and dust. Buy the best quality feed and hay you can afford within your budget. Because prices and nutritional requirements vary so much from one geographical location to another, consult goat keepers in your area or your local agricultural extension office for advice.

Risk Management

One important business expense is to purchase insurance to cover your facilities and your animals. Farm insurance policies cover theft, loss, and damage to your facilities, equipment, vehicles, and even your home and its contents. Livestock insurance covers the cost of replacing livestock. Liability insurance covers accidental injury to employees or visitors to your farm. Talk to your insurance agent, agricultural extension office, or lender about the amount of coverage you need.

You can protect your income by purchasing federal Adjusted Gross Revenue-Lite (AGR-Lite) insurance (**www.agrlite.net**), a whole-farm plan that protects farmers against low revenue due to unavoidable natural disasters and market fluctuations. Most farm-raised crops, animals, and animal products are eligible for protection. To apply for AGR-Lite you will need to provide:

- A history calculation worksheet, including five years of allowable income and expense data from IRS tax returns (Schedule F or equivalent forms)

- An annual farm report for the insurance year listing each commodity to be produced, the expected quantity of the commodity to be produced, and the expected price for the commodity

Operating a dairy goat business is not the same as keeping dairy goats as a hobby, to provide milk for your family, or to diversify your farm. You must locate a reliable market for your products and provide a steady supply of milk. The income from the sale of your products should be enough to pay the costs of managing and feeding your goats and justify the effort it takes to keep the business going. Milk production is the first priority; milk production levels have to be closely watched and unproductive does culled from the herd. The larger your herd, the greater the risks you undertake.

Fortunately, many resources are available to assist you in making a business plan and carrying it out, including academic studies, reference books on dairy farming, agricultural extension officers, and the advice of local goat keepers. The more research you do before you start your business, the greater your chances of success.

oats are small, versatile, adaptable, and easy to accommodate. Dairy goats are an excellent addition to a family farm. In addition to providing milk and manure and keeping pastures clear of brush, they often become family pets. Their friendly and sociable nature makes them easy for children to handle. As a commercial business, dairy goat farming is experiencing unprecedented growth in the United States due to a growing interest in natural foods and an expanding market for ethnic cuisines. Now is an ideal time to get started in the dairy goat business.

After reading this book, you should be confident that you have the knowledge to acquire your own herd, set up a shelter and milking facilities, and manage your goats successfully. What you do not know, you will soon learn from your local agricultural extension office, your veterinarian, and the farmers in your community. Goat owners are enthusiastic about their animals and willing to share their expertise with anyone interested.

As you have learned from the chapters on feeding, nutrition, and goat diseases, your goats will thrive as long as they are provided with the right feed and a reasonably clean living environment. There is a vast library of scientific literature and educational publications about every aspect of raising and managing goats. *In Appendix C, you will find a list of resources where you can get more information on a variety of subjects relating to dairy goats.*

May you and your goats enjoy long and happy years of companionship!

Sample Budget Worksheet for a Dairy Goat Business

Dairy Goats 100 Head Unit
Class #2 Grade Herd, Per Doe Basis

Operating Inputs	Units	Price	Quantity	Value (price X quantity)
Mixed feed	CWT		7.200	_____
Alfalfa hay	Tons		0.900	_____
Vet medicine	Herd		1.000	_____
Supplies	Herd		1.000	_____
Utilities	Herd		1.000	_____
Doe replacement feed	Herd		1.000	_____
Kid feed	Herd		1.000	_____
Breeding fees	Herd		1.000	_____
Miscellaneous expense	Herd		1.000	_____

Operating Inputs	Units	Price	Quantity	Value (price X quantity)
Marketing expense	Herd		1.750	_____
Machinery labor	Hr.		0.847	_____
Equipment labor	Hr.		1.630	_____
Livestock labor	Hr.		7.692	_____
Machinery fuel, lube, repairs	DOL		1.000	_____
Equipment fuel, lube, repairs	DOL		1.000	_____
Total operating costs				0.00

Fixed Costs	Interest rate on loans or credit	Cost	Value	Total Value
Machinery				
Interest at depreciation, taxes, insurance		(Cost X interest)		_____
Equipment				
Interest at depreciation, taxes, insurance		(Cost X interest)		_____
Livestock				
Doe goat			_____	_____
Buck goat			_____	
Replacement doe goat			_____	
Interest at depreciation, taxes, insurance		(Cost X Interest)		_____
Total fixed costs				0.00

Production	Units	Price	Quantity	Value	Your Value
Goat milk	CWT		20.00		_____
Male kids	Herd		0.90		_____
Female kids	Herd		0.65		_____
Cull doe goats	Herd		0.20		_____
Total receipts					

Returns above total operating cost	_____
Returns above all specified costs	_____
5% doe death loss, 200% kid crop	
10% kid death loss, 25% doe replacement rate	

Developed and processed by Department of Agricultural Economics, Oklahoma State University

Sample Goat Sale Contract

se this contract as a template and select the terms and conditions appropriate for your circumstances. Many goat breeders have posted their sales contracts online, and you may find additional terms you wish to add.

Seller's name: _____

Mailing address: _____

City, state & zip: _____

Phone: _____

Cell phone or fax: _____

E-mail: _____

Buyer's name: _____

Mailing address: _____

City, state & zip: _____

Phone: _____

Cell phone or fax: _____

E-mail: _____

This contract is for the sale of the following goat(s), herein referred to as "the goat," to (buyer's name)_____, herein referred to as the "Buyer," by (seller's name)_____, herein referred to as the "Seller," for the total sum following after each of the goat's names:

_____ ID _____ $_____

_____ ID _____ $_____

_____ ID _____ $_____

TOTAL $_____

Seller's Disclaimer and Buyer's Responsibilities

Seller is responsible for and will:

- Feed, house, and provide basic maintenance, health care, and worming to the goat until the agreed upon pickup date unless other arrangements are made.

- Give possession of the goat to the buyer on the agreed upon date after full payment is received.

- Provide all applicable registration certificates or registration applications to Buyer within fifteen (15) days of Seller's receipt of full payment.

Buyer is responsible for and will:

- Pick up the goat on or before_____.

- Pay the full purchase price with cash, cashier's check, or money order at or by the time of pickup, or with a personal check at least ten (10) days prior to pickup.

- Pay for and provide adequate transportation of the goat from or to the facilities of Seller and to or from the Buyer's facilities. All goats must be transported in an adequately sized, covered, ventilated, safe vehicle where they may comfortably and safely stand, turn around, and lie down. If the Seller deems transportation unsatisfactory/unsafe, Seller reserves the right to refuse to release the goat.

- Pay for Veterinary Health Certificate (optional, but required by law for interstate travel), and fees charged by the vet for the farm call and animal examination and tests.

Terms and Conditions:

- The down payment/deposit is nonrefundable. If the buyer decides to cancel the purchase, the down payment/deposit will not be refunded.

- The goat must be picked up and paid in full by _____(DATE). If the goat is not picked up and paid in full by this date, $5.00 will be charged every day after (DATE). If the goat is not picked up within a week of the agreed upon date, the buyer forfeits the animal as well as the monies, unless other arrangements have been agreed upon in writing.

- Goats are sold "as is." Seller guarantees that at the time of sale, goat(s) are in good health, free of injury or disease and that the goat(s) will breed if provided with the proper care, environment, and nutrition. Seller makes no other guarantees or warranties, either expressed or implied. Any subsequent claims by Buyer, contesting Seller's representation as to the health, physical condition, or breeding soundness of the goat(s) at date of sale must be fully substantiated by a physical examination and applicable medical tests performed by a licensed veterinarian and provided in writing to the Seller.

- In the unlikely event that a goat should die, become ill, or miscarry, while still in the possession of Seller, the goat will be replaced with one of equal value as soon as one is available or a full refund will be given.

- Buyer agrees to provide adequate food, care, shelter, and fresh water for the goat(s) at all times. It is not acceptable to keep a goat tied out or on a chain. If at any time it is found that these basic needs are not being met, the authorities will be notified, and Seller will resume ownership of the goats. No refunds will be given.

The deposit/down payment is $_____, due on the date this contract is signed. By paying a deposit/down payment, the Buyer agrees to, and is bound by, all terms, conditions, and responsibilities laid out in this contract. By signing this contract, the Buyer agrees that he or she has read and understood this entire contract and agrees to all the responsibilities, terms, and conditions laid out herein.

Accepted and agreed to this (DATE)_____,

Signed _____ , Buyer

Printed Name _____

Signed _____ , Seller

Printed Name _____

Helpful Websites

Feeding Goats

Goat Ration Calculator. Langston University
(**www.luresext.edu/goats/research/rationbalancer.htm**)

List of Plants Poisonous to Goats
(**www.ars.usda.gov/Main/docs.htm?docid=10086**)

List of Plants Poisonous to Goats
(**http://fiascofarm.com/goats/poisonousplants.htm**)

List of Plants Poisonous to Goats
(**http://netvet.wustl.edu/species/goats/goatpois.txt**)

Ration Evaluator Excel Spreadsheet
(**www.sheepandgoat.com/Spreadsheets/RationEvaluator2004.xls**)

Ration Mixer Excel Spreadsheet
(**www.sheepandgoat.com/spreadsheets/rationmixer.xls**)

Rhododendron poisoning antidote
(**www.goatworld.com/health/plants/antidotes.shtml**)

Rumen Physiology and Rumination. University of Colorado
(**www.vivo.colostate.edu/hbooks/pathphys/digestion/
herbivores/rumination.html**)

The Small Ruminant Nutrition System
(**http://nutritionmodels.tamu.edu/srns.htm**)

Goat's milk

Cheesemaking recipes
(**www.agmrc.org/media/cms/zeng04_A2E10C94923B2.pdf**)

Composition of goat's milk, courtesy of the USDA
(**www.everything-goat-milk.com/goat-milk-table.html**)

Composition of goat's milk
(**www.dairyforall.com/goatmilk-composition.php**)

Federal requirements for processing milk
(**www.federalregister.gov/articles/2010/10/05/2010-24985/
milk-for-manufacturing-purposes-and-its-production-and-
processing-requirements-recommended-for#p-3**)

Goat Milk Versus Cow Milk
(**www.goatworld.com/articles/goatmilk/goatmilk.shtml**)

Information about milking dairy goats
(**http://fiascofarm.com/goats/milking.htm**)

Raw milk laws by state
(**www.realmilk.com/happening.html**)

The World's Healthiest Foods: Goat's Milk
(**www.whfoods.com/genpage.
php?tname=foodspice&dbid=131**)

Dairy goat business references

"Dairy Goats: Sustainable Production." *Livestock Production
Guide.* Linda Coffey, Margo Hale, and Paul Williams. NCAT
Agriculture Specialists. August 2004. ATTRA Publication #IP258
(**http://attra.ncat.org/attra-pub/dairygoats.html**)

General information about dairy goats

Beginners Guide to Dairy Goats. American Goat Society
(**www.americangoatsociety.com/registration/pdf/
BeginnersGuidetoDairyGoats.pdf**)

Best Management Practices for Dairy Goat Farmers. Wisconsin
Dairy Goat Association
(**www.wdga.org/resources/bmp8.pdf**)

Dairy Goat Production Guide. University of Florida
IFAS Extension. Barnet Harris Jr. and Frederick Springer. 2009.
(**http://edis.ifas.ufl.edu/ds134**)

Goat anatomy diagrams
(**www.dpi.nsw.gov.au/__data/assets/pdf_file/0010/178336/goat-
anatomy.pdf**)

Ireland Agricultural and Food Development Authority
(**www.teagasc.ie/ruraldev/progs/goats/**)

USDA Cooperative Extension System Offices
(www.nifa.usda.gov/Extension/index.html)

Breeding and raising kids

American Dairy Goat Association (ADGA) genetics website
(www.adgagenetics.org)

Goat Identification Used by Registries. Goat ID.
NAIS Working Group
(http://usanimalid.com/registryID.htm)

National Animal Identification System (NAIS)
(www.dhia.org/06%20Feb%20AIN_Admin.pdf)

Video on bottle-feeding a kid
(www.righthealth.com/topic/bottle_feeding/Anatomy)

Organizations

Alpines International Breed Club
(www.alpinesinternationalclub.com)

American Dairy Goat Association
(www.adga.org)

American Goat Federation
(www.americangoatfederation.org)

American Goat Society
(www.americangoatsociety.com)

American LaMancha Club
(www.lamanchas.com)

International Nubian Breeders Association
(**www.i-n-b-a.org**)

Kinder Goat Breeders Association
(**www.kindergoatbreeders.com/resources.html**)

Maryland Small Ruminant Page
(**www.sheepandgoat.com**)

National Association of Dairy Regulatory Officials (NADRO)
(**www.nadro.org**)

Nigerian Dwarf Goat Association (NDGA)
(**www.ndga.org**)

National Institute of Animal Agriculture (NIAA)
(**www.animalagriculture.org**)

National Pygmy Goat Association
(**www.npga-pygmy.com**)

National Saanen Breeders Association
(**http://nationalsaanenbreeders.com**)

National Toggenburg Club
(**http://nationaltoggclub.org**)

Oberhasli Breeders of America
(**http://oberhasli.net**)

USDA Animal and Plant Inspection Services Animal ID
(**www.aphis.usda.gov/traceability/downloads/NAIS-UserGuide.pdf**)

Plans

Buck goat yard
(**http://bioengr.ag.utk.edu/extension/extpubs/Plans/6300.pdf**)

"Construction of High Tensile Wire Fences." R. A. Bucklin, W.
E. Kunkle, and R. S. Sand. Document CIR851 of the Agricultural
and Biological Engineering Department, Florida Cooperative
Extension Service, Institute of Food and Agricultural Sciences,
University of Florida
(**http://edis.ifas.ufl.edu/ae017**)

Goat shelters
(**www.goatworld.com/articles/shelters_gwmf.shtml**)

"Grounding Electric Fences." Tom Cadwallader and Dennis
Cosgrove. University of Wisconsin-Extension
(**www2.uwrf.edu/grazing/ground.pdf**)

Metal goat milking stand
(**http://bioengr.ag.utk.edu/extension/extpubs/Plans/6399.pdf**)

Mineral feeder
(**http://bioengr.ag.utk.edu/extension/extpubs/Plans/5916.pdf**)

Plans for a milking barn and milk house for ten dairy goats
(**http://bioengr.ag.utk.edu/extension/extpubs/Plans/6255.pdf**)
(**http://bioengr.ag.utk.edu/extension/extpubs/Plans/6256.pdf**)

Standalone Hay and Grain Feeder
(**http://bioengr.ag.utk.edu/extension/extpubs/Plans/5910.pdf**)

Supplies

Fly predators
(**www.spalding-labs.com/Dairy/Default.aspx**)

Hamby Dairy Supply
(**http://hambydairysupply.com**)

Hartford Livestock Insurance
(**www.hartfordlivestock.com**)

Information on a commercial dairy goat business

"2005 Sample Costs for a 500 Dairy Goat Operation." University of California Cooperative Extension
(**http://coststudies.ucdavis.edu/files/dairygoatsnc05r.pdf**)

"Agricultural Alternatives: Dairy Goat Production." Penn State University. 2008.
(**http://agalternatives.aers.psu.edu/Publications/dairy_goat. pdf**)

"Costs and Returns for Dairy Goat, 1500 lbs. Milk/Doe, 100 Doe Herd." Rutgers University, Table 79
(**http://aesop.rutgers.edu/~farmmgmt/ne-budgets/organic/ DAIRY-GOAT-1500LB-MILK.HTML**)

"Dairy Goats and a Sustainable Future." Rona Sullivan
(**www.dairygoatjournal.com/issues/83/83-6/Rona_Sullivan. html**)

Dairy Goat Glossary

abomasum. The last of the four compartments in a goat's digestive system, where true digestion occurs

ADGA. American Dairy Goat Association

advanced registry. A doe documented to have given a set amount of milk over the course of a year (high-yield)

alcilli. Another name for alveoli

alveoli. Tiny sack-like structures containing cells that secrete milk in a mammary gland

American. A doe that is seven-eighths pure of one breed or a buck that is fifteen-sixteenths pure

anthelmintics. Medications that eliminate worms

artificial insemination (AI). The impregnation of a doe using frozen sperm from a donor buck

blind teat. A nonfunctional teat

broken-mouthed. A term for a goat missing some of its front incisors

buck. A male goat

buckling. An uncastrated male kid

Burdizzo. A tool used to neuter young bucks

CAE. Caprine arthritic encephalitis, a viral rheumatic joint inflammation passed from mother to kid

castrate. To render a buck sterile

chamois. A deep, reddish-brown color found in the Oberhasli breed

chevon. Goat meat

chèvre. A type of goat cheese originating in France

colostrum. The first nourishment a mother provides to her kid

concentrate. Feed that contains a high proportion of nutrients and is low in crude fiber content (less than 18 percent of dry matter)

confinement feeding. Feeding goats a controlled diet inside a shelter or yard

conformation. The degree to which an individual goat matches the ideal standard for its breed

cud. Soft masses of undigested plant fiber that return from the rumen to the mouth for additional chewing

deep littering. Allowing a mattress of bedding to accumulate on the floor of the stable by adding fresh hay on top every day

dental palate. A hard, tough pad of tissue that takes the place of teeth in a goat's upper jaw

disbudding. The process of cauterizing a kid's horn buds

doe. A female goat

doeling. A young doe that has not mated for the first time

dual-purpose breed. A breed of goat that can belong to more than one of the three usage categories: milk, meat, or fiber producing

ENE. Estimated net energy, provided by carbohydrates and fats

elastrator. A device used to castrate a kid using a tight rubber ring

elf ear. The short, two-inch ear of a LaMancha goat

estrone sulfate. A hormone produced by a living fetus about 35 days after conception

estrus. The state in which a goat's ovary contains a fertile egg, and her uterus is ready to establish it

experimental. A goat resulting from an accidental breeding between purebred goats of different breeds

flehmen. The raising of the head and drawing up of the upper lip by a buck that is ready to mate

floppy kid syndrome (FKS). A disease that affects kids between 3 and 10 days old. Symptoms are muscular weakness and depression, progressing to flaccid paralysis, and often death.

foot rot. A bacterial infection that causes liquification of tissue in and around the hoof

foot scald. A red, moist bacterial infection of the tissue between the toes

freshening. The commencement of milk production after kidding

Furstenberg's rosette. A many-folded mucus membrane that prevents leakage of milk from an udder and acts as a barrier to bacteria

gopher ear. The almost non-existent ear of a LaMancha goat

gummer. A goat that has lost all its teeth

hardware disease. Life-threatening damage to the wall of the reticulum, caused when a goat swallows sharp or pointed objects

hermaphroditism. The presence of both male and female reproductive organs

hybrid vigor. The tendency of a mixed-breed animal to be healthier and stronger than a purebred

ketosis. A disease of pregnant goats occurring when the body breaks down its own reserves of fat

kid. Newborn or immature goat

kidding. The process of giving birth

lactation. The production of milk

let down. The release of milk within the udder prior to milking

line breeding. The breeding of closely related goats to intensify their genetic traits

loose stable. A stable in which the goats are housed together in an open area instead of in individual stalls

lungworm. A parasite that infests the lungs and respiratory passages

mammary gland. The part of the udder that secretes milk

mastitis. A common inflammation of the udder generally caused by poor sanitation practices, insect bites becoming infected, or injury to low hanging udders

meconium. The first feces passed by a newborn kid

monkey mouth. A condition in which the lower front teeth protrude beyond the edge of the upper dental palate

NASS. USDA's National Agricultural Statistics Service

N-P-K (nitrogen-phosphorus-potassium) value. A measure of the nutrients contained in a fertilizer

National Animal Identification System (NAIS). A program to assign identification to each farm and its livestock so diseased livestock can be quickly tracked

native on appearance (NOA). A goat that appears to be of a specific breed but is undocumented

omasum. The third of the four compartments in a goat's digestive system. Its function is to absorb nutrients.

oxytocin. A hormone that controls milk let down

pannier. A basket worn by a pack goat

papered. Registered with a breed association

parrot mouth. A condition in which the lower incisors fall behind the leading edge of the upper dental palate

pasteurization. The process of heating milk to a specific temperature and maintaining it for a specific length of time, in order to kill harmful bacteria

pedigree. A document or chart with the recorded ancestry of a goat, its parents, grandparents, and so on

placenta. Afterbirth

polled. A goat that is born hornless

purebred. Goats that, according to ancestry, fall into a breed group defined by national and often international breed standards

recorded grades. Goats that meet a list of requirements

regarding appearance and quality of milk

registered purebred. A goat that has been registered with an official registry organization and is listed in the herd book

registration. Documents verifying that an animal is registered in the official herd book of a recognized registry organization for that breed

reticulum. The second of the four compartments in a goat's digestive system, which serves as a fluid pump

roughage. Highly fibrous plant material, feed with a fiber content higher than 18 percent

rumen. The largest of the four compartments in a goat's digestive system, where fiber is broken down

ruminant. An animal whose digestive system consists of multiple stomachs

sausage teats. Teats that are exceptionally wide

scurs. Misshapen horns that can twist down into the goat's face

settle. To become pregnant

soiling. Confinement feeding

sow mouth. A condition in which the lower front teeth protrude beyond the edge of the upper dental palate

stanchion. A head restraint used to hold a goat during milking, hoof trimming and inspections

star milker. A system exists whereby a goat is given a star rating depending upon her consistent yield (measured day by day) and the percent of butterfat her milk contains. A star milker generally must give 10 to 11 pounds of milk in a day.

stays. The horizontal wires on a wire fence that hold the vertical wires in place

step-in fence posts. Posts made of metal, plastic, or fiberglass that can be inserted into the ground using body weight

TDN. Total digestible nutrients, provided by carbohydrates and fats in a goat's diet

T-posts. Metal fence posts used to fill in the spaces between wooden corner posts

teat. The opening through which milk comes out of the udder

teat dip. A disinfectant applied to teats after each milking to prevent mastitis

topline. The line of a goat's spine

udder. The part of the goat containing the mammary glands that produce milk

udder attachments. The ligaments that attach the udder to the goat's abdominal wall

udder supports. Another name for udder attachments

urinary calculi. Bladder stones

wattle. A fold of skin under the face on the side of the neck

wether. A castrated buck

Belanger, Jerome D. *Storey's Guide to Raising Dairy Goats*. Pownal, Vermont: Storey Books, 2001.

Coffey, Linda, Margo Hale, and Paul Williams, NCAT Agriculture Specialists. "Dairy Goats: Sustainable Production." *Livestock Production Guide* (August 2004). Accessed March 16, 2011. **http:// attra.ncat.org/attra-pub/dairygoats.html#resources**.

Jaudas, Ulrich, Fritz W. Kohler, and Matthew M. Vriends. *The Goat Handbook*. New York: Barron's, 2006.

Luttmann, Gail. *Raising Milk Goats Successfully*. Charlotte, Vermont: Williamson Pub., 1986.

North, Robert. "Anatomy and Physiology of the Goat." NSW Department of Primary Industries. Agfact A7.0.3, second edition (2004) **www.dpi.nsw.gov.au/__data/assets/pdf_file/0010/178336/goat-anatomy.pdf**.

Smith, Cheryl K. *Raising Goats for Dummies*. Hoboken, New Jersey: Wiley, 2010.

USDA – NASS Wisconsin Field Office. *Sheep and Goat Review. Volume 1, Issue 1*. (February 2010).

Weaver, Sue. *Goats: Small-scale Herding for Pleasure and Profit*. Hobby Farm Press/BowTie Press, 2006.

artha Maeda has lived and worked in Australia, Japan, Latin America, and several African countries, where she experienced firsthand the importance and versatility of goats in providing meat and milk even in the harshest conditions. She currently lives with her family in Orlando, Florida.

Index